UNIVERSITY
LIBRARY
NOTTINGHAM

Occupational Ergonomics

Work related musculoskeletal disorders are a leading cause of worker impairment, disability, compensation costs and loss of productivity in industrialized countries. The ageing of the workforce and presence of physically demanding jobs contribute to the widespread occurrence of musculoskeletal disorders.

This edited volume presents an overview of the critical issues related to musculoskeletal disorders at work and brings together the latest research in this field. Each of the contributors is an internationally renowned expert in their own right.

The book will be of interest to a wide range of professionals and researchers in ergonomics, occupational health, epidemiology, psychology and engineering. It will also serve as an important source of information for policy makers.

Francesco Violante is currently Director of the Regional Program in Occupational Health, a centre for research, teaching and clinical investigation into occupational disorders in Bologna, Italy. **Thomas Armstrong** is the Director of the Center for Ergonomics at the University of Michigan, USA and **Åsa Kilbom** is at the National Institute for Working Life, Solna, Sweden.

Occupational Ergonomics

Work related musculoskeletal disorders of the upper limb and back

Edited by Francesco Violante,
Thomas Armstrong and
Åsa Kilbom

London and New York

First published 2000 by Taylor & Francis
11 New Fetter Lane, London EC4P 4EE

Simultaneously published in the USA and Canada
by Taylor & Francis Inc.
29 West 35th Street, New York, NY 10001

Taylor & Francis is an imprint of the Taylor & Francis Group

This book has been produced from camera-ready copy provided by the editors

Printed and bound in Great Britain by MPG Books Ltd, Bodmin, Cornwall

British Library Cataloguing in Publication Data
A catalogue record for this book is available
from the British Library

Library of Congress Cataloging in Publication Data
Occupational ergonomics : work related musculoskeletal disorders of the upper limb and back / [edited by] Francesco Violante, Åsa Kilbom, and Thomas Armstrong.
p.; cm.
Includes bibliographical references and index.
ISBN 0-7484-0933-5
1. Musculoskeletal system – Diseases. 2. Musculoskeletal system – Wounds and injuries. 3. Industrial accidents. 4. Human engineering. I. Violante, Francesco. II. Kilbom, Åsa. III. Armstrong, Thomas J.
[DNLM: 1. Occupational Diseases – prevention & control – Congresses. 2. Arm Injuries – prevention & control – Congresses. 3. Back Injuries – prevention & control – Congresses. 4. Human Engineering – Congresses. 5. Neck Injuries – prevention & control – Congresses. 6. Occupational Health – Congresses. WA 440 O145 2001]
RC925.5 . O26 2001
616.7′ – dc21

ISBN 0 7484 0933 5

Contents

Preface by Elisabeth Lagerlöf

This book is dedicated to Professor Åsa Kilbom, MD: it is a tribute to her vast scientific contribution to field of ergonomics and to her job in organising a course which originated this work. She unfortunately fell ill right before the course started. We all wish her a good recovery!

SIGNIFICANT ILL HEALTH PROBLEM

There is substantial evidence within the EU member states and the US that neck and upper limb musculoskeletal disorders and back disorders are a significant problem with respect to ill health and associated costs within the workplace. It is likely that the size of the problem will increase because workers are becoming more exposed to work risk factors for these disorders within the EU.

Therefore, there is a need for a book that can serve as an introduction to and a useful reference for those who are working in the field of ergonomics. This is why these proceedings have been published thanks to the support of Taylor & Francis Ltd. The target group of the book are practitioners within the occupational health services and students who need to get an overview of the whole field of ergonomics.

The importance of the musculoskeletal problem is, for instance, also confirmed by a recent review on *Work-related neck and upper limb musculoskeletal disorders* published by the European Agency for Safety and Health at Work in Bilbao, Spain. The European Commission (DG V) had requested the assistance of the Agency to conduct a review of the available scientific knowledge regarding risk factors for work-related neck and upper limb musculoskeletal disorders. The Agency invited Professor Peter Buckle and Dr Jason Devereux of the Robens Centre for Health Ergonomics, UK to facilitate the study and prepare the report.

Combating musculoskeletal disorders is also a top health and safety priority for the European Trade Union Confederation. The ETUC Executives Committee decided in 1997 to start a European-wide campaign on this issue. A booklet with initiatives taken in this effort was recently published by the European Trade Union Technical Bureau for Health and Safety Europe under strain. A report on trade union initiatives to combat workplace musculoskeletal disorders by Rory O Neill.

NOT A NEW PROBLEM

Musculoskeletal disorders is no recent problem. Already in 1706 Bernardo Ramazzini, an Italian physician considered as the father of occupational health, wrote about office work:

> The diseases ... arise from three causes; first constant sitting, the perpetual motion of the hand in the same manner, and thirdly the attention and the application of the mind. Constant writing also considerably fatigues the hand and the whole arm on account of the continual and almost tense tension of the muscles and tendons.

Today, 30 percent of all workers report that they suffer from back pain and 17 from the upper limbs according to the Second European Survey on Working Conditions in 1996; 45 percent report that they are working in tiring or awkward positions. The compensated musculoskeletal disorders vary widely between countries due to different diagnosis criteria and different compensation systems.

Work-related musculoskeletal disorders arise when exposed to work activities and work conditions that significantly contribute to their development or exacerbation, but not acting as the sole determinant of causation (World Health Organisation, 1985).

Estimates of the costs are limited and where data exists, i.e. in the Nordic countries and The Netherlands, the costs have been estimated at between 0.5% and 2% of the GNP. The costs for work-related musculoskeletal disorders in the US have been estimated at $20 billion per year and the indirect costs at around $60 billion per year. There is, however, a lack of standardised assessment criteria, which makes comparisons between countries difficult.

A number of epidemiological studies have found that women are at a higher risk for work-related upper neck and limb disorders although associations with workplace risk factors generally are found to be stronger than gender factors. The importance of gender differences and their implications for work system design requires more substantial debate.

The Agency report also pointed out the need of specific and sensitive diagnostic criteria for work-related musculoskeletal disorders. Whilst this is recognised, the expert panel suggested that in general the prevention strategies recommended or put in practice to avoid the risks of these disorders would not be dependent upon the diagnostic classification. All

MSDs, even those without a specific diagnosis or pathology, should be considered in health monitoring and surveillance systems.

NIVA FOR EDUCATION AND TRAINING

NIVA, the Nordic Institute for Advanced Training and Education in Occupational Health, started with courses in ergonomics already in the late 1980s. The problem was early recognised in the Nordic countries and a vast amount of research was emerging as well as an ever increasing absence from work due to musculoskeletal disorders. The courses started in order to disseminate the latest research results and to discuss prevention strategies.

NIVA is funded by the Nordic Council of Ministers and educates researchers, experts and qualified practitioners about the risks in and challenges of working life. The course language is English. Information technology is used for strategic and long-term improvements of NIVA activities.

The NIVA concept for education and training in safety and health, and other work life matters consists of

- Advanced courses based on Nordic and international front knowledge
- Target groups: Researchers and practitioners
- 3–5 day courses in English, 10–12 per year
- Development of courses with the aid of a scientific network
- Course leaders from and courses held in all Nordic countries
- Internationally well-known lecturers
- Good reputation for Nordic cooperation among NCM organisations

NIVA courses provide an excellent opportunity to meet and build a network with colleagues from other Nordic countries and abroad. The working language of NIVA is English in order to form a basis for the Nordic and international collaboration and also to make it possible to accept non-Nordic qualified applicants.

EUROPEAN COLLABORATION IMPORTANT

NIVA has arranged 1–2 courses every year in ergonomics. Over a period of time, more and more participants from Europe and outside Europe have attended the courses.

In 1995 Finland and Sweden joined the European Union. The Nordic Council of Ministers has since then encouraged more European activities in parallel with the Nordic ones and those directed towards the Nordic immediate surroundings, i.e. the Baltic states and the St Petersburg area.

It was therefore of great pleasure to me and to the NIVA Board to accept the proposal Dr Francesco Violante made early in March 1998 about a joint course in Ergonomics in Italy under the NIVA umbrella. In Italy, musculoskeletal disorders – other than vibration white fingers, have only been compensated in the last 2–3 years. The claims for disorders have however increased sharply during this time from 893 reports in 1996 to 2000 in 1999.

Funds were also made available to support the course by the Emilia Romagna Regional Department of Health and University of Bologna, Italy.

A SUCCESSFUL COURSE

The course proved to be a success. More than 80 persons wanted to participate in the course but it was only possible to accept about 60 of whom 70% were from the Nordic countries and 30% from the rest of Europe, mostly Italians. About one-third came from research institutions, one-third from occupational health services and the rest from health institutions, authorities and other companies.

The course, like these proceedings, started with epidemiology of the back, neck and upper limb disorders presented by Dr Hillka Riihimki, Finland and Dr Mats Hagberg, Sweden.

Upper limb risk factors were introduced by Dr Tom Amstrong, USA, and vibration and musculoskeletal disorders by Dr Massimo Bovenzi, Italy. Dr Thöres Teorell, Sweden, approached the psychosocial risk factors.

Screening for musculoskeletal disorders was presented by Dr Alf Franzblau, USA, while Dr Stover Snook, USA was responsible for back risk factors. Dr Francesco Violante, Italy, introduced case definitions for the musculoskeletal disorders.

Dr Don Chaffin, USA, introduced biomechanics, and Dr Daniela Colombini talked about redesigning jobs, while Dr Georgio Aresini, European Commission, participated among the others in a panel debate on regulatory issues on occupational ergonomics.

The aim of the course was accomplished in that it provided the participants with both research and how to work in practice. Åsa Kilbom's sudden illness forced some last minute changes in the course framework, but the other lecturers quickly changed their agenda in order to cover as much as possible of her lectures. The Italian participants in particular found the course very useful because of the stress on diagnosis criteria, while some of the Nordic participants would have liked more attention paid to prevention strategies.

I want in particular to thank Dr Francesco Violante, who has carried a heavy burden both during the course and in the preparation of the proceedings (with the tireless help of Dr Francesca Graziosi). His engagement and that of his co-workers made the course so successful, and we are looking forward to further collaboration.

Editors and contributors

EDITORS

Francesco Violante, Occupational Health Unit, Sant'Orsola Malpighi Hospital, Bologna, Italy, EU.

Thomas J. Armstrong, Center for Ergonomics, University of Michigan, Ann Arbor, MI, USA.

Åsa Kilbom, National Institute for Working Life, Solna, Sweden.

CONTRIBUTORS

Giorgio A. Aresini, European Commission, Lusembourg, EU.

Renzo Bergamaschi, Embraco, Riva di Chieri, Turin, Italy, EU.

Massimo Bovenzi, Occupational Health Institute, Cancer Center, Trieste, Italy, EU.

Don B. Chaffin, Center for Ergonomics, University of Michigan, Ann Arbor, MI, USA.

Daniela Colombini, Center for Occupational Medicine, Milan, Italy, EU.

Alfred Franzblau, School of Public Health, University of Michigan, Ann Arbor, MI, USA.

Candido Girola, Embraco, Riva di Chieri, Turin, Italy, EU.

Mats Hagberg, Occupational Medicine, Sahlgrenska University Hospital, Gothenburg, Sweden, EU.

Lucia Isolani, School of Occupational Medicine, University of Bologna, Bologna, Italy, EU.

Elisabeth Lagerlöf, Director, NIVA, Helsinki, Finland, EU.

Enrico Occhipinti, Center for Occupational Medicine, Milan, Italy, EU.

Maurice Oxenburgh, Lilyfield, NSW, Australia.

Giovanni Battista Raffi, School of Occupational Medicine, University of Bologna, Bologna, Italy, EU.

Hilkka Riihimäki, Finnish Institute of Occupational Health, Helsinki, Finland, EU.

Stover H. Snook, Harvard School of Public Health, Boston, MA, USA.

Töres Theorell, Department of Occupational Health, National Institute of Psychosocial Factors and Health, Stockholm, Sweden, EU.

Chapter 1

Methodological issues in epidemiologic studies of musculoskeletal disorders

Hilkka Riihimäki

1.1 INTRODUCTION

Epidemiology is defined as the study of the distribution and determinants of health-related states and events, and the application of this study to control of health problems (Last 1995). Descriptive epidemiology provides valuable data for the administration and health care providers to plan and implement health policy and to evaluate its effectiveness. Etiologic epidemiology aims at identifying the risk factors of diseases and provides the basis for the action to prevent the diseases by reducing or eliminating exposure to the risk factors. The third type is prognostic (or intervention or experimental) epidemiology in which the effectiveness of interventions (change of exposure, interventions to alter the course of a disease) is studied.

Compared to epidemiologic research on some other chronic diseases, such as cancer or coronary heart disease, epidemiology of musculoskeletal disorders has relatively short tradition. Interest in this area of research has increased only during the past twenty years, although some of the classic studies date back to the 1950s. One explanation is that only during the past 20 years musculoskeletal diseases have started to be considered a major public health problem. Another possible issue is that a researcher in this field is facing some particularly difficult problems. In this chapter some basic methodological issues of epidemiologic research of musculoskeletal disorders are discussed. For more in-depth reading on epidemiologic methods, textbooks such as the one by Rothman and Greenland (1998) are recommended.

1.2 TYPES OF EPIDEMIOLOGIC STUDY

1.2.1 Cross-sectional study

Most epidemiologic information about musculoskeletal disorders is based on cross-sectional studies. In a cross-sectional study, disease and exposure information is ascertained simultaneously. Often there is selection among the cases of the disease, because in a cross-sectional setting cases with long duration are overrepresented. For a current case of the disease, the current exposure may not be relevant; in etiologic considerations only exposures before the onset of the disease are of importance. Cross-sectional surveys serve administrative purposes well, but they are less suitable for etiologic studies. Among the advantages of cross-sectional studies is that they are relatively easy, quick and inexpensive to perform. Major disadvantages are selection bias (e.g. selection among cases or 'healthy worker effect') and the difficulty to discern the temporal sequence between the exposure and health outcome.

1.2.2 Cohort study

In a cohort study, the cohort of exposed and unexposed people is formed from the population that does not have the disease of interest, i.e. the 'healthy' population. The cohorts are followed up over time and the incidence of new cases of the disease is recorded (Figure 1.1).

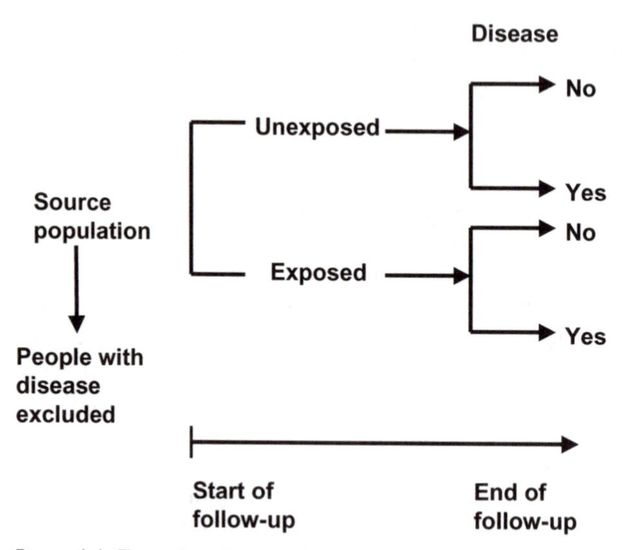

Figure 1.1 Flow of a cohort study.

Unlike many chronic diseases with clear-cut diagnostic criteria, for musculoskeletal disorders it is often difficult to define the healthy population and often no attention has been paid to this matter in epidemiologic studies of musculoskeletal disorders. For example, for intermittently occurring low back pain it is difficult to know whether the spells of low back pain represent acute phases in a continuum of a disease process or whether the spells are independent attacks of injury. The researcher should make some assumptions about the nature of the disease and, accordingly, make an operational definition for the 'healthy' population to be followed up. Cohort studies with careful collection of data on exposure, confounding, and modifying factors and the health outcome are least susceptible to bias of all types of epidemiologic studies. A cohort study allows multiple outcomes to be studied simultaneously and for rare exposures it is the only feasible option.

Cohort studies take a long time to carry out, and therefore they are usually expensive. Cohort studies are suitable only for diseases that are common enough in the population.

1.2.3 Case-control study

In a case-control study, first the cases of the disease under study are defined and identified. Any definition can be used for a case. The case definition determines the source population of the cases and then the controls can be drawn from this source population. It is not necessary to include all possible cases, but the cases can be randomly selected. It is important that the sampling is independent of the exposure of interest. The same holds true for the sampling of the controls. Data of exposure and other determinants of the disease are obtained retrospectively for the cases and controls (Figure 1.2).

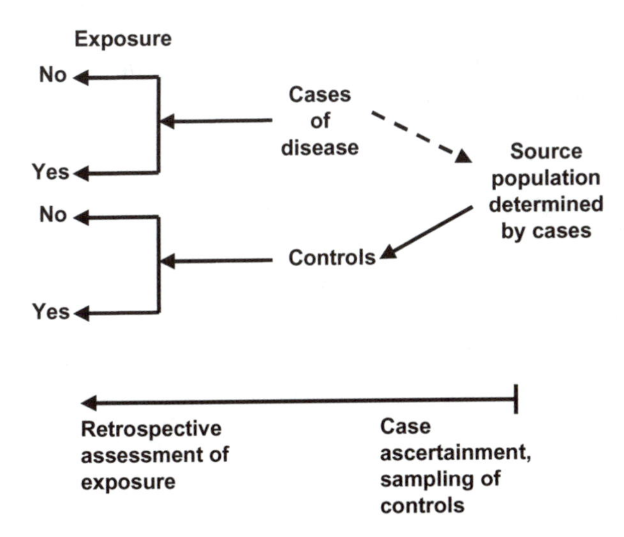

Figure 1.2 Flow of a case-control study.

Sometimes a case-control study can be embedded into a cohort study: cases are identified among the members of the cohort and either noncases or a sample of them are used as controls. In such 'nested' case-control studies the quality of exposure data is usually better than in other types of case-control studies.

In general, case-control studies are less expensive to carry out than cohort studies. It is possible to study the effects of many exposures at the same time, and a case control study is the only feasible option for rare diseases. Case-control studies are more susceptible to bias than cohort studies. A particular source of bias is the potential misclassification of exposure and other determinants of the disease due to the retrospective assessment. Another problem in the field of musculoskeletal disorder research is the difficulty in case detection.

1.2.4 Case-crossover study

A special case of case-control study is case-crossover study, in which each case serves as its own control: for each case one or more earlier time periods are selected as matched control periods. The exposure status of a case at the time of the disease onset is compared with the distribution of exposure status for that same individual in earlier time periods. Case-crossover study is suitable to investigate triggering effects, such as the causes of accidental injuries. Case-crossover study is quick and easy to perform. The most important limitation of this study type is that only exposures varying within individuals can be studied. This design is suitable only to study exposures with a short induction time and transient effects. Case-crossover design has not been extensively used, but in the domain of musculoskeletal disorders the question of the triggering factors of symptoms or injuries is a highly relevant one.

1.3 ETIOLOGIC AND PROGNOSTIC RESEARCH

In the course of a disease, there are two distinctly different phases of interest from the epidemiologic point of view, the one before and the one after the onset of the disease. In the former phase, the interest is in the etiology or risk factors of the disease and in the latter in the prognosis or factors affecting the future course of the disease (Figure 1.3).

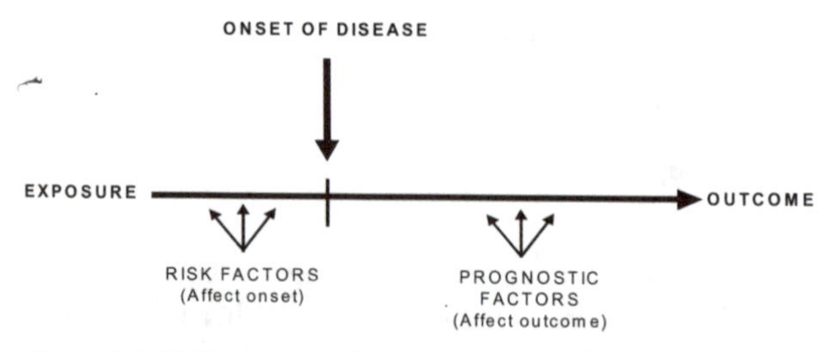

Figure 1.3 Differentiation between risk factors and prognostic factors.

This basic dichotomy is often neglected in epidemiologic studies of musculoskeletal disorders. In etiologic studies only exposures that have occurred before the onset of the disease matter, and likewise, in prognostic studies only exposures or interventions, such as ergonomic changes, treatment or rehabilitation that take place after the disease onset are relevant. This division is important because, for instance, work-related etiologic and prognostic factors may differ from each other. In a cross-sectional setting these two phases of the disease cannot be disentangled. The negligence of paying attention to their differentiation is easy to understand, because for some musculoskeletal disorders the time of onset is difficult to determine.

The definition of health outcome in epidemiologic studies varies according to the objectives of the study. In etiologic studies in which the health outcome of interest has an acute nature (like acute low back pain), first-time incident cases should be recorded in a healthy population. If the health outcome has a chronic nature, such as osteoarthritis, prevalent cases can be used as the health outcome in an etiologic study, but the onset should be estimated as well as possible. In case of osteoarthritis, the onset could be defined as the time when the first symptoms occurred.

In prognostic studies, in which the objective is either to study the natural course of the disease and its determinants or the effect of various interventions (ergonomic changes, treatment or rehabilitation, etc.), a definition is needed for a case of the disease as well as for the cure or exacerbation of the disease. It is important that all cases that are included in the study are in the same phase in the course of the disease. For instance, if the prognosis of acute low back pain is being studied, the duration and type of symptoms and preferably also the past history of low back pain of the subjects should be similar.

1.4 MUSCULOSKELETAL DISORDER CASE ASCERTAINMENT

A common means to identify cases of chronic diseases in epidemiologic studies is the use of register data. In the best national cancer registers, for instance, all cases of cancer are being registered from different sources of information. Such comprehensive register data are seldom available for musculoskeletal disorders. Data from some registers, such as hospital discharge registers, pension registers, or registers of health care services can be used, but these data often carry bias. The reporting of a disease case to the register may depend on exposure status; a person with heavy physical demands in his/her job may have to go on pension whereas a person in a lighter job with a musculoskeletal disorder of the same severity may continue working. Such selection can lead to spurious associations between exposures and the health outcome. Another problem in using register data is nonuniformity among the medical profession in using the diagnostic labels for musculoskeletal disorders. The diagnostic entities vary from one country to another and from one doctor to another, even though there are international classification schemes, such as the one by the WHO (WHO 1992).

Often the most feasible option to catch the musculoskeletal disorder cases is a survey, which means that all subjects in the study are systematically examined. As an example, low back disorder case ascertainment can be based on questionnaire information about symptoms as well as the findings of physical examination and function tests or imaging.

As to the cause of the current episode, the stage of the disorder can be classified to acute or chronic and sometimes also a category for subacute is used. No generally accepted classification criteria for acute and chronic low back disorders exist, but three-months' duration of symptoms is a common cut-off point. In population studies of low back disorders physical examination and function tests have not proved to provide much useful information for the classification of the disorders and, thus, in most studies only low back pain has been assessed. For many other musculoskeletal disorders findings in clinical examination (e.g. epicondylitis) or clinical examination and function tests (carpal tunnel syndrome), or in imaging (osteoarthritis) together with symptoms are required for diagnoses. It is important that in epidemiologic studies clear-cut definitions for diagnoses or syndromes are given.

Using symptoms to ascertain the health outcome is inexpensive and data acquisition is quick and easy (postal questionnaires or interviews). It is particularly suitable for large studies and, in principle, standardization is easy. One of the most commonly used standardized symptom questionnaires is the 'Nordic Questionnaire' (Kuorinka et al. 1987). It suits surveillance purposes well, but further development of good standardized methods to ascertain symptom-based musculoskeletal syndromes are needed. Another matter of concern with the use of symptom-based disorder ascertainment is that symptoms are always a subjective perception. This perception can be influenced by several extraneous factors, such as other concomitant diseases, physical activity level at work and leisure, individual characteristics, such as pain and illness behavior or mental state, and also by cultural factors. This raises questions about the validity of symptom-based classification. Further validity considerations are caused by recall error in symptoms reporting and also by the deficient knowledge about the relationship between symptoms and pathological processes in the musculoskeletal system. One further problem with low back pain, in particular, is that simple low back pain is so common in the population that it is difficult to define the 'healthy' subpopulation.

Epidemiology of symptom-based musculoskeletal disorders could be improved by improving the questionnaires, and classification of the disorders and by validating the classifications. Operational case definitions can be based on pain localization, pain characteristics (intensity, type), the duration of the pain, the frequency of pain episodes, and the occurrence of other symptoms. In questionnaires and interviews, special attention should be paid to verbal expression and differences in dialects or cultures. Other means to assess symptoms include visual analogue scales and pain drawings.

Imaging reveals structural changes or derangements which can be associated with pathologic processes in the musculoskeletal system. The most common objects of imaging studies are osteoarthritis (X-ray, MRI), intervertebral disc degeneration (X-ray, MRI, CT), but also soft tissue disorders such as tendon disorders (MRI, ultrasonography) or bursitis (US, MRI) can be studied. Imaging provides objective data, but the evaluators need to be carefully blinded with regard to the subject characteristics. If clinical imaging material is used for case-referent studies, there may be selection bias due to differential referral of patients to medical care according to their exposure status. The disadvantages of imaging include high expenses (MRI), radiation exposure (X-ray, CT), requirement of highly qualified experts in the research team (MRI, US), and also validity aspects are of concern. The reliability of the evaluation of the images should be assured as well as the relevance of the imaging findings.

1.5 EXPOSURE ASSESSMENT

As mentioned before, in etiologic studies only exposure preceding the onset or inception of the disease is relevant. Depending on the etiopathomechanism of the disease, the exposure of interest may by the current, recent or cumulative past exposure. The time from the start of exposure to the inception of the disease is called induction time and the time from the inception to the detection of the disease, latency (Rothman and Greenland 1998) (Figure 1.4).

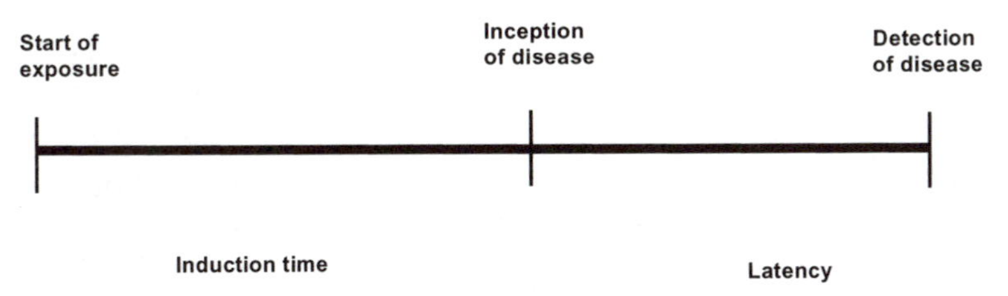

Figure 1.4 Definition of induction time and latency.

In practice, often these two periods cannot be defined separately, but they are combined together and called the latency period. Latency has seldom been taken into consideration in epidemiologic studies of musculoskeletal disorders. It is, however, most important to make some assumptions of the induction and latency periods, if the true periods are unknown. Such assumptions help in planning the exposure assessment strategy. Different assumptions can also be applied in the analysis phase of the study, provided exposure data are available for different time periods.

The relationship between physical load and musculoskeletal disorders is commonly described with a U-shaped curve: too little as well as too much load increase the risk of musculoskeletal disorders (Winkel and Westgaard 1992). It is undoubtedly true if immobilization is included in the scale of physical load. However, if the scale is restricted to normal activity range only, evidence from epidemiologic studies does not give strong support to the model (Bernard et al. 1997).

Hazardous exposure to physical load can be conceptualized as overload. Overload can be caused by several mechanisms: sudden overexertion that causes injury; sustained or repetitive exertion that can cause fatigue or injury. Operationally exposure to different modes of physical load, external load, posture and motion, is assessed. Each of them can be classified according to frequency of occurrence, duration, and intensity (or amplitude). Particularly in large epidemiologic studies it is often not possible to measure accurately load, posture and motion. In many cases it would suffice to have a valid classification scheme to categorize exposure to four classes: neutral exposure (not fatiguing or injurious to the musculoskeletal system), exposure to low, medium and high overload.

For appropriate exposure assessment, the relevant time or time window to assess the exposure should be hypothesized. This should be guided by the knowledge or assumptions about the induction and latency times of the disease. Also the knowledge of the etiopathomechanism of the disease should guide the selection of the relevant type of exposure measure. What is the interesting characteristic of loading? Is it the mean level or the peaks? Of the peaks, is it the highest peak ever or regularly occurring peaks? Is it sufficient to estimate the time-weighted average or is the cumulative exposure before the onset of the disease more important? All measures of physical load can be transformed into measures of force and, accordingly, the use of a common metric of force has been proposed (Walsh et al. 1997).

Exposure can be assessed via self reports, observation methods or direct measurements. Today sophisticated technology is available to carry out accurate continuous measurements of physical load even at work sites. As described by Winkel and Mathiassen (1994), cost and exactness increase from self reports to observations methods to direct measurements. There is a reversed direction of change for capacity, versatility and generality (Figure 1.5).

Thus the selection of the assessment method depends on the goals and setting of the study as well as on economic and practical feasibility. One particular problem with measurement data is how to reduce the large

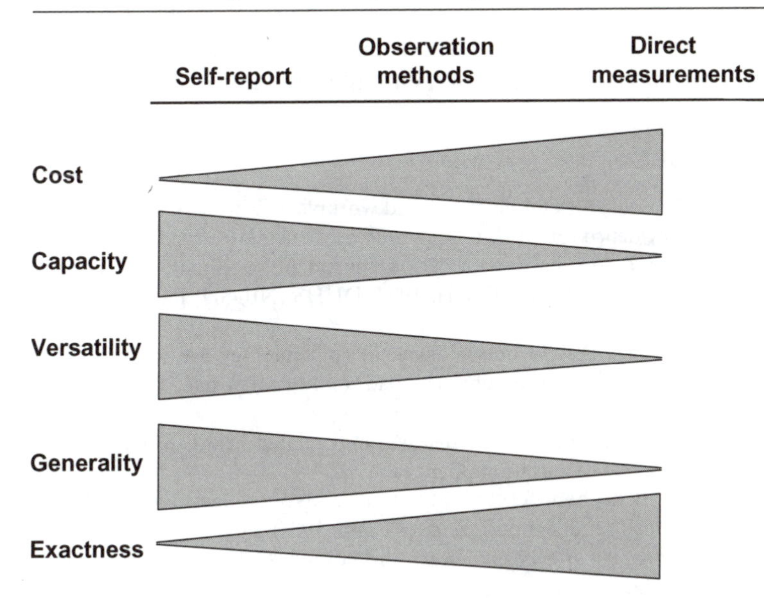

Figure 1.5 Differences between exposure assessment methods: self-reports, observation methods and direct measurements (adapted from Winkel and Mathiassen 1994).

amount of data into some meaningful measures. It is good to bear in mind a statement by Lotfi Zadeh, one of the first developers of fuzzy logic: 'As complexity rises, precise statements lose meaning and meaningful statements lose precision' (McNeill and Freiberger 1973).

Self-reports of exposure are easy to obtain even in large population studies. Torgén et al. (1999) have studied the validity and reliability of retrospective self reports by comparing them with observation and self reported data collected on the same persons six years earlier. Validity was fair to good for perceived exertion, sitting, kneeling and squatting, and repetitive work; it was poor for head bent forward, head rotation, and trunk bent forward. Reliability was found to be fair to good for perceived exertion, sitting, kneeling and squatting, head rotation, hands above shoulder, repetitive finger movements, and manual materials handling; it was poor for head bent forward, trunk bent forward, trunk rotation, and repetitive hand movements. It has commonly been suspected that there is differential misclassification in self-reported exposure to physical load between those with musculoskeletal disorders and those without. In the study by Torgén et al., no indication of such differential misclassification was detected.

The results of the study by Torgén et al. indicate that self reports are in many occasions a sufficiently valid exposure assessment method for epidemiologic studies. In planning the questionnaires the researchers should, however, keep in mind two pieces of good advice given by Burdorf and van der Beek (1999): Do not ask questions that cannot be answered and do not ask questions for which the answers are already known.

1.6 REFERENCES

Bernard BP (ed.). Musculoskeletal disorders and workplace factors. A critical review of epidemiologic evidence for work-related musculoskeletal disorders of the neck, upper extremity, and low back pain. US Department of Health and Human Services CDC (NIOSH), Cincinnati, OH, 1997: DHHS (NIOSH) Publication No. 97-141.

Burdorf A, van der Beek AJ. In musculoskeletal epidemiology are we asking the unanswerable in questionnaires on physical load? (editorial) *Scand J Work Environ Health* 1999; 25: 81–3.

International statistical classification of diseases and related health problem, tenth edition. World Health Organisation, Geneva, 1992.

Kuorinka I, Jonsson B, Kilbom Å, et al. Standardised Nordic questionnaire for the analysis of musculoskeletal symptoms. *Appl Ergon* 1987; 18: 233–7.

Last JM (ed.) *A Dictionary of Epidemiology*, third edition. Oxford University Press, New York, NY, 1995.

McNeill D, Freiberger P. *Fuzzy Logic*. Simon & Schuster, New York, NY, 1993: 43.

Rothman KJ, Greenland S. *Modern Epidemiology*. Lippincott-Raven Publishers, Philadelphia, PA, 1998.

Torgén M, Winkel J, Alfredsson L, Kilbom Å. Evaluation of questionnaire-based information on previous physical work loads. *Scand J Work Environ Health* 1999; 25: 246–54.

Walsh R, Norman R, Neuman P et al. Assessment of physical load in epidemiologic studies: common measurement metrics for exposure assessment. *Ergonomics* 1997; 40: 51–61.

Winkel J, Mathiassen SE. Assessment of physical load in epidemiologic studies: concepts, issues, and operational considerations. *Ergonomics* 1994; 37: 979–88.

Winkel J, Westgaard RH. Occupational and individual risk factors for shoulder–neck complaints. Part III the scientific basis (literature review) for the guide. *Int J Ind Ergon* 1992; 10: 85–104.

Chapter 2

Epidemiology of work-related back disorders

Hilkka Riihimäki

2.1 INTRODUCTION

The prevalence estimates of back disorders vary depending on assessment methods and population characteristics. Most commonly described entity is low back pain (LBP), which is intermittent in nature. For such an entity, instead of point prevalence, a period prevalence, i.e. the proportion of the population experiencing low back pain during a certain period of time, is often presented. The period of observation can vary from lifetime to the past year or one month. Most data on back disorders are based on questionnaire surveys. Only in a few community studies has clinical examination been used to verify the existence of back disorders. On the other hand, it is a common understanding that with a thorough clinical examination some rare, specific causes of back pain, such as malignancies, infections, inflammatory diseases, or fractures can be identified. In addition to that, only a differentiation between sciatica and nonspecific back symptoms can be done (Bigos et al. 1994).

In a community study in the U.K. (Bradford), the prevalence of LBP was investigated in the 25–64-year-old population (Hillman et al. 1996). The lifetime prevalence of LBP was 59%, 12-month prevalence 39%, and point prevalence 19%. Over a one-year period, LBP was acute (duration of symptoms less than 2 weeks) in 50%, subacute (duration more than 2 weeks but less than 3 months) in 21%, and chronic (duration at least 3 months) in 26% of the cases. Time off work due to LBP had been taken by 6%.

In Table 2.1, age-adjusted prevalence rates of back disorders are presented for the Finnish women and men at least 30 years of age (Heliövaara et al. 1993). The data are from a representative sample of the population and the diagnoses were based on the history of low back disorders, symptoms, and physicians' examinations. There was no significant difference between men and women, except for point prevalence of clinically verified sciatica or herniated disc that was higher among men (5.1%) than among women (3.7%).

Table 2.1 Back disorders in the Finnish population over 30 years of age. Age-adjusted prevalence rates (%) (Heliövaara et al. 1993).

	Women	*Men*
Lifetime prevalence of back pain	73.3	76.3
Lifetime prevalence of sciatic pain	38.8	34.6
Five-year prevalence of sciatic pain requiring bedrest for at least two weeks	19.4	17.3
One-month prevalence of low back or sciatic pain	23.3	19.4
Point prevalence of clinically verified low back pain syndrome	16.3	17.5
Sciatica or herniated disc	3.7	5.1*

*p = 0.005.

Figure 2.1 shows the age-dependency of the prevalence of back pain, defined in various ways in two general-population-based surveys, the Mini Finland Health Survey (Heliövaara et al. 1993) and the National Health and Nutrition Examination Survey (Deyo and Tsui-Wu 1987). According to both surveys, the prevalence of LBP increases with increasing age until 55–64 years and after that there is a decline. The figure also depicts the great influence of different definitions of LBP on the prevalence estimates.

Hildebrandt (1995) has reported the prevalence of back pain in the Dutch working population according to trades and professions. In men the professions with the highest prevalence rates were construction workers, cleaners, supervisory production workers, plumbers and drivers; the professions with the lowest prevalence rates were chemists, scientists, bookkeepers, and civil servants. In women, the profession with the highest prevalence rate was cleaners and those with the lowest rates were secretaries and 'other services'.

Only little information is available about the incidence rate of low back pain. In Bradford community, the annual incidence was 4.7% (Hillman et al. 1996). In a prospective cohort study of three occupational groups with distinctly different exposure to physical load at work, a clear-cut difference was detected in the incidence of first-time occurrence of sciatic pain (low back pain radiating to the leg) according to work-related exposure to physical load. The three-year cumulative incidence was 24% among construction carpenters exposed to dynamic physical work, 22% among machine operators exposed to whole body vibration and prolonged sitting in a constrained posture, and 14% among office workers exposed to sedentary work (Riihimäki et al. 1994).

Sciatic pain and local low back pain seem to represent, at least partly, different phases of low back disorder. In a study of Finnish forest industry employees, a hierarchical classification of sciatic pain, local low back pain and no low back pain was used. A further classification criteria was the duration of the symptoms. Local low back pain of short duration (less than two weeks) was most common in the youngest age groups whereas

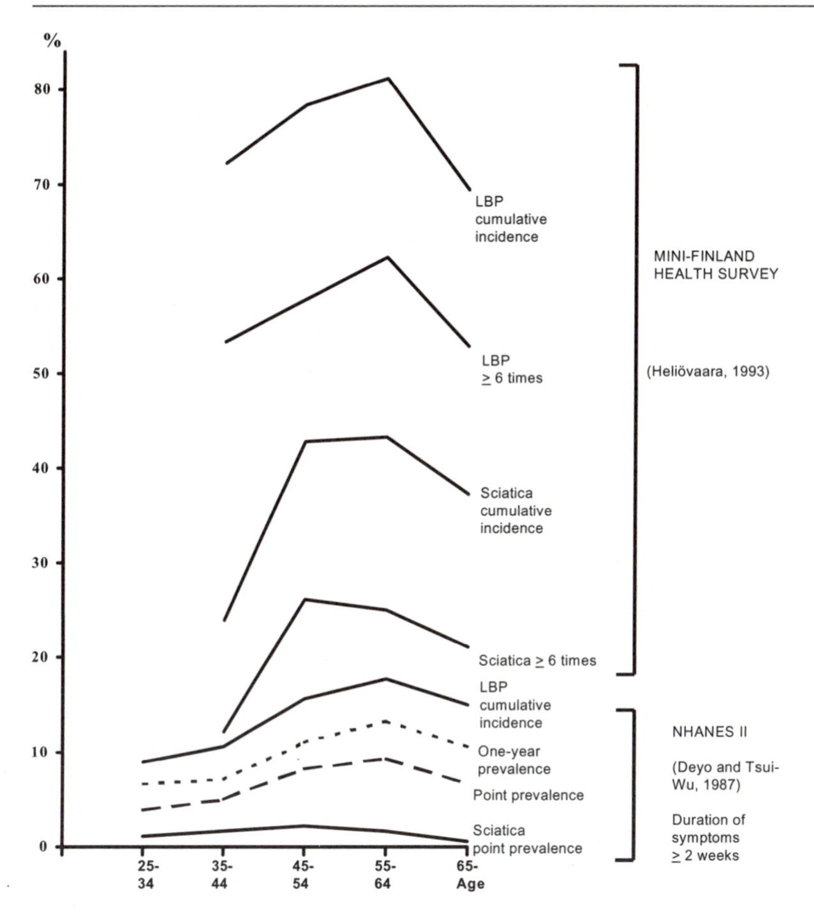

Figure 2.1 Prevalence rates of low back pain by age in two general population surveys.

sciatic pain with long duration (more than two weeks) increased in prevalence towards the older age groups. This pattern was similar for both the lifetime and 12-month prevalences. (Riihimäki 1995).

A common notion is that low back pain has become more common in the industrialized countries during the past twenty years. There is little information about the changes in the prevalence rates of low back pain or low back syndromes but the notion is mostly based on the statistics of injury claims or disability, which depict not only the morbidity but also the social consequences caused by the morbidity. The National Institute of Public Health in Finland has carried out an annual questionnaire survey on a random sample of the population since 1985. The time series have not shown any increase in the prevalence of low back pain but rather a slight decrease has been demonstrated (Leino et al. 1994). A similar trend has been reported by the

Statistics Finland; the prevalence of recurrent lumbo-sacral pain among employees in Finland has decreased from 27% in 1977 to 25% in 1997 among women and from 29% to 26% among men, respectively (Lehto and Sutela 1999).

In most cases low back pain has a favorable course but recurrences are common. It has been estimated that within one month since the onset of low back pain, 35% become symptom free and return to normal activity. After 3 months 87% and after 12 months 96% have become symptom free, but during the 12 months 40–50% have contracted recurrent symptoms (Frymoyer and Howe 1984). Many studies have shown similar patterns for healing times (Andersson 1999). The intermittent nature of the symptoms is one factor that makes epidemiologic and intervention research on low back pain a challenging task.

Recent studies have shown that the natural course of low back pain is not always so favorable. According to a study by von Korff and Saunders (1996), after a one-year follow-up of primary care back patients, one-third of them had intermittent or persistent pain, one of seven had pain with severe intensity, and one of five had substantial activity limitations. Wahlgren et al. (1997) followed people with their first onset of low back pain that had occurred on a daily basis for the previous 6–10 weeks. After 12-months' follow-up, 72% continued experiencing the pain and 14% had marked disability.

It is important to take notice of the type of symptoms when considering the prognosis of LBP. In the cohort of Finnish forest industry employees, the most probable symptom 'pathways' of the future occurrence of local and radiating LBP was scrutinized during a three-year follow-up using repeated questionnaires (Riihimäki et al. 1998). Among those without any LBP during the previous 12 months at the baseline, the most probable path was to remain symptomfree throughout the whole follow-up period (38%) and the most probable type of pain to appear was local LBP. The most persistent symptom was concomitant radiating and local LBP: almost half of those with such symptoms at the baseline reported the same combination of symptoms throughout the three-year follow up. From both the health care and epidemiologic research point of view it is important to gain better knowledge about the natural course of LBP and the factors affecting it.

2.2 WORK-RELATED RISK FACTORS OF LOW BACK PAIN

Much literature is available about work-related risk factors of low back pain. Most studies are cross-sectional with inherent sources of bias which weaken the quality of evidence. Recently several comprehensive reviews of the available evidence have been published.

The National Institute of Occupational Safety and Health (NIOSH) in the USA included in its review of musculoskeletal diseases and workplace factors all studies which had at least 70% participation rate, in which health outcome was defined by symptoms and clinical examination (this criterion was not strictly applied for low back disorders), investigators were blinded for health outcome/exposure assessment, and the joint (anatomic region) under study was subjected to independent exposure assessment (Bernard 1997). Of a total of over 2000 studies, a thorough review of evidence was done from more than 600 studies. For back disorders about 40 studies were considered as providing evidence. Of physical factors, strongest evidence was found for lifting and forceful movements whereas the evidence was insufficient for static postures. Of psychosocial factors, the evidence was not considered as strong for any of the risk factors (Table 2.2).

Table 2.2 Evidence of work-related risk factors of low back pain according to a review by NIOSH (Bernard et al. 1997).

Risk factor at work	*Strength of evidence*
Physical factors	
heavy physical work	evidence
lifting and forceful movements	strong evidence
whole-body vibration	strong evidence
static postures	insufficient evidence
Psychosocial risk factor	
intensified work load	suggested
job dissatisfaction	mixed
low job control	limited support
monotonous work	mixed
social support	weak evidence

Burdorf and Sorock (1997) investigated positive and negative evidence of risk factors for back disorders using different eligibility criteria for studies. Their eligibility criteria were: relative risk was or could be estimated; study population was not a patient population; and participation rate was more than 50%. They identified 35 studies from the past 15 years that fulfilled these criteria. Most convincing evidence was found for manual materials handling, whole-body vibration, and frequent bending and twisting as being risk factors of back disorders. The evidence was not convincing for static work postures and repetitive motions. The evidence was mixed also for psychological factors, stress, job dissatisfaction, work pace and job decision latitude or monotonous work. No association was found for lack of job support.

The most rigorous review was reported by Hoogendoorn et al. in 1999. Their study was based on a strictly systematic approach to find and summarize the evidence. They included only cohort (28 studies) and case-referent (3) studies, cross-sectional studies were excluded. According to their review, strong evidence exists for manual materials handling, bending and twisting, and whole-body vibration as risk factors for back pain. Moderate evidence was found for patient handling and heavy physical work, and no evidence for standing or walking, sitting, sports, and total leisure-time physical activity. The three reviews show quite a consistent picture which indicates that potentially there should be a good chance to reduce the risk of back disorders at the work place by reducing the exposure to physical risk factors. Evidence for the effectiveness of ergonomic interventions is, however, sparse and particularly in the USA there is an ongoing debate about this matter. It is a very demanding task to carry out a properly controlled ergonomic intervention study, but that should be one of the high priorities for the future research.

2.3 DISC DEGENERATION

Population data about the prevalence of disc degeneration of the lumbar spine date back to the 1950s and 1960s when Lawrence performed the classic radiographic studies in the UK (Lawrence 1969). They showed that the prevalence of lumbar disc degeneration increases steeply with increasing age and that disc degeneration is more prevalent among men than women (Figure 2.2).

Lawrence (1955) also showed that radiographically detected lumbar disc degeneration is associated with occupational load: miners and heavy manual workers were more affected than light manual workers or sedentary workers, whether the severity of disc degeneration or the number of intervertebral spaces affected was considered.

Only indirect information of advanced disc degeneration can be detected using radiography. Nowadays a much more sensitive technique, magnetic resonance imaging (MRI) is in use also in epidemiologic studies. One of the advantages of MRI compared to X-rays is that MRI does not have any known adverse health effects. An MRI study of middle-aged (40–45 years) construction carpenters, machine operators and office workers indicated that the prevalence of decreased signal intensity of the nucleus pulposus, a sign of general disc degeneration, is not related to occupational load, but posterior bulges of the discs were most common among the carpenters and anterior disc bulges among the machine operators (Luoma et al. 1998).

There is increasing evidence that disc degeneration is quite common among adolescents. In a three-year follow up of 15-year-old school children with low back pain and their symptom-free controls it was seen that disc degeneration of the lumbar spine was significantly more common among the cases than controls. At the end of the follow up the prevalence of disc degeneration was 58% among the cases and 26% among the referents (Salminen et al. 1995).

Battié and coworkers (1995) have carried out a twin study which showed

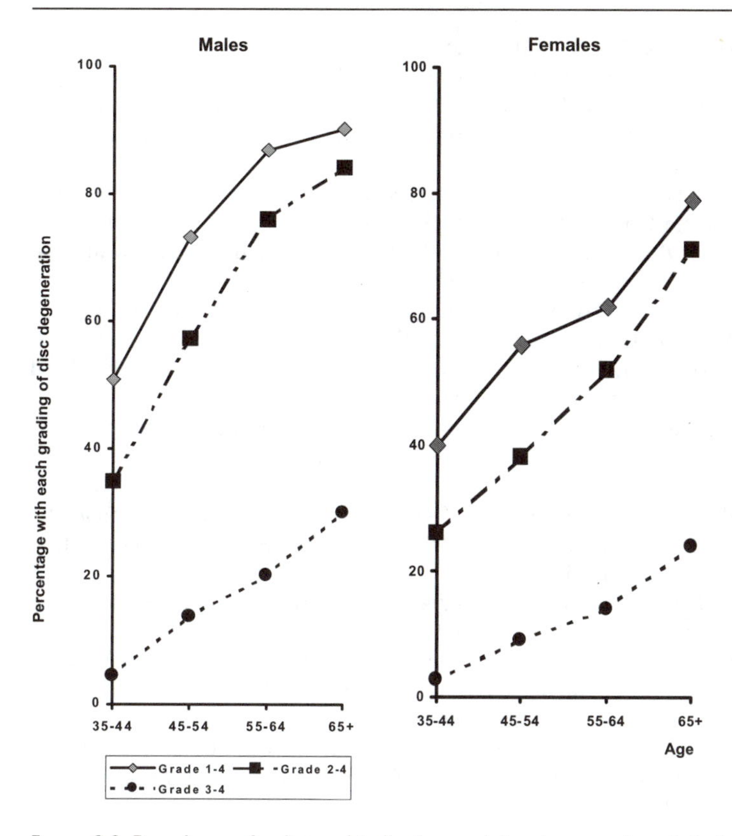

Figure 2.2 Prevalence of radiographically detected disc degeneration of the lumbar spine by age and sex (Lawrence 1969).

that familial aggregation explains a much greater proportion of the variability in lumbar disc degeneration than physical loading. There is also evidence that the polymorphism of vitamin D receptor gene and some collagen genes is associated with susceptibility to disc degeneration (Videman et al. 1998, Kawaguchi et al. 1999). Future studies are needed to learn about the interactions between the genetically determined susceptibility and environmental factors, such as physical load in the etiopathogenesis of disc degeneration.

2.4 REFERENCES

Andersson BGJ. Epidemiological features of chronic low-back pain. *Lancet* 1999; 354: 581–5.

Battié MC, Videman T, Gibbons LE, Fisher LD, Manninen H, Gill K. Determinants of lumbar disc degeneration: a study relating lifetime exposures and magnetic resonance imaging findings in identical twins. *Spine* 1995; 20: 2601–12.

Bernard BP, ed. Musculoskeletal disorders and workplace factors. A critical review of epidemiologic evidence for work-related musculoskeletal disorders of the neck, upper extremity, and low back pain. U.S. Department of Health and Human Services CDC(NIOSH), Cincinnati, OH, 1997: DHHS(NIOSH) Publication No. 97-141.

Bigos S, Bowyer O, Braen G et al. Acute low back problems in adults. Clinical practice guideline No. 14. AHCPR Publication No. 95-0642. Agency for Health Care Policy and Research, Public Health Service, U.S. Department of Health and Human Services, Rockville, MD, December 1994.

Burdorf A, Sorock G. Positive and negative evidence of risk factors for back disorders. *Scand J Work Environ Health* 1997; 23: 243–56.

Deyo RA, Tsui-Wu Y-J. Descriptive epidemiology of low-back pain and its related medical care in the United States. *Spine* 1987; 12: 2648.

Frymoyer JF, Howe J. Clinical classification. In: Pope MH, Frymoyer JF, Andersson G (eds.). *Occupational Low Back Pain.* Praeger Publishers, New York, 1984: 71–98.

Heliövaara M, Mäkelä M, Sievers K. *Tuki- ja liikuntaelinten sairaudet Suomessa* [Musculoskeletal diseases in Finland]. Publication of the Social Insurance Institution, Helsinki, 1993: AL:35.

Hildebrandt VH. Back pain in the working population: prevalence rates in Dutch trades and professions. *Ergonomics* 1995; 38: 1283–98.

Hillman M, Wright A, Rajaratnam G, Tennant A, Chamberlain MA. Prevalence of low back pain in the community: implications for service provision in Bradford, UK. *J Epidemiol Community Health* 1996; 50: 347–52.

Hoogendoorn WE, van Poppel MNM, Bongers PM, Koes BW, Bouter LM. Physical load during work and leisure time as risk factors for back pain. *Scand J Work Environ Health* 1999; 25: 387–403.

Kawaguchi Y, Osada R, Kanamori M et al. Association between an aggregan gene polymorphism and lumbar disc degeneration. *Spine* 1999; 24: 2456–60.

Lawrence JS. Disc degeneration. Its frequency and relationship with symptoms. *Ann Rheum Dis* 1969; 28: 121–38.

Lawrence JS. Rheumatism in coal miners: Part III Occupational factors. *Br J Ind Med* 1955; 12: 249–61.

Lehto A-M, Sutela H. Efficient, more efficient, exhausted. Findings of Finnish quality of worklife surveys 1977–1997. Statistics Finland, Helsinki, Labour Market, 1999: 8.

Leino P, Berg M-A, Puska P. Is back pain increasing? Results from national surveys in Finland during 1978/9-1992. *Scand J Rheumatol* 1994; 23: 269–76.

Luoma K, Riihimäki H, Raininko R, Luukkonen R, Lamminen A, Viikari-Juntura E. Lumbar disc degeneration in relation to occupation. *Scand J Work Environ Health* 1998; 24: 358–66.

Riihimäki H. Defining outcome in epidemiologic studies of low back pain. In Nordman H, Starck J, Tossavainen A, Viikari-Juntura E (eds.) Proceedings of the sixth FIOH–NIOSH joint symposium on occupational health and safety Finnish Institute of Occupational Health, Helsinki, People and work Research reports No. 3, 1995: 155–8.

Riihimäki H, Viikari-Juntura E, Moneta G, Kuha J, Videman T, Tola S. Incidence of sciatic pain among men in machine operating, dynamic physical work, and sedentary work – a three-year follow-up. *Spine* 1994; 19: 138–42.

Riihimäki H, Viikari-Juntura E, Martikainen R, Takala E-P, Malmivaara A. *Book of Abstracts*. International Society for the Study of the Lumbar Spine, Brussels, 9–13 June 1998: 141.

Salminen JJ, Erkintalo M, Laine M, Pentti J. Low back pain in the young. A prospective three-year follow-up study of subjects with and without low back pain. *Spine* 1995; 20: 2101–7.

Videman T, Leppävuori J, Kaprio J, Battié MC, Gibbons LE, Peltonen L, Koskenvuo M, Intragenic polymorphisms of the vitamin D reseptor gene associated with intervertebral disc degeneration. *Spine* 1998; 23: 2477–85.

Von Korff M, Saunders K. The course of back pain in primary care. *Spine* 1996; 21: 2833–9.

Wahlgren DR, Atkinson JH, Epping-Jordan JE, et al. One-year follow-up of first onset low back pain. *Pain* 1997; 73: 213–21.

Chapter 3

Epidemiology of neck and upper limb disorders and work place factors

Mats Hagberg

3.1 WHY IS THE EPIDEMIOLOGY IMPORTANT?

Disorders of the neck and upper limb are common problems in the general population as well as among industrial workers. In the general working population in Sweden, as many as one-third of women and one-quarter of men reported pain in the neck and shoulder that was present every day or every other day. Knowledge about the epidemiology of neck and upper limb disorders is important for different types of prevention as well as for handling medical legal issues. In primary prevention we need to know the risk factors for neck and upper limb disorders to design workplaces and work systems that promote health for the worker. By using epidemiological methods we can identify the risk factors and their magnitude that can be used for prioritization of where initiation and implementation of change at work is most needed. In secondary prevention (treatment of the injured worker to full recovery) involves early workplace rehabilitation where knowledge of the prognosis of different neck and upper limb disorders is important. To accommodate the injured worker with impaired function at the workplace knowledge of factors that prevents disability is important. Much effort over the last twenty years has been devoted to the medical legal aspects of neck and upper limb disorders and workplace factors. The debate and statements made both generally and in individual patient cases have been fogged by improper use of/or misinterpretation of epidemiological data.

3.2 WHAT ARE THE NECK AND UPPER LIMB DISORDERS?

A patient with a neck and upper limb disorder, evaluated by two physicians may get two different diagnoses (Hagberg 1996). There is a lack of consensus not only concerning diagnostic criteria but also on examination techniques for the neck and upper limb disorders (Buchbinder et al. 1996). Furthermore there is also a discrepancy between epidemiologists, ergonomists and clinicians in the use of terminology for neck and upper limb disorders.

3.2.1 Non-specific musculoskeletal pain

The most common type of neck and upper limb disorders in the Scandinavian countries is pain in the neck/shoulder region with or without neck stiffness and with tenderness over the descending part of the trapezius muscle (non-specific neck/shoulder pain). In epidemiological literature this is sometimes called tension neck syndrome (Viikari-Juntura 1983) (Waris et al. 1979), but clinicians seldom use this type of diagnosis. Among physicians there is also common with 'misclassification', a patient with non-specific neck/shoulder pain is diagnosed as a myalgia patient although proper diagnostic techniques have not been used, i.e. muscle biopsy to ascertain that the pain originates from the muscles. Rather this should be called, according to ICD-10 (International Classification of Diseases WHO), a cervical brachial pain syndrome (M53.1). It is also common with non-specific pain in the forearm and in the wrist and hand (Ranney et al. 1995).

3.2.2 Tendinitis

The second most common disorder in the working population is probably tendinitis. Tendinitis is inflammation of the muscle tendon, the attachment of the muscle to the bone. The inflammation may not cause cellular infiltration and other mechanism than inflammation may cause the pain (Moore 1992). Common locations for tendinitis are the shoulder (rotator cuff-, supraspinate-, biceps-tendinitis), the elbow as a lateral epicondylitis or to the wrist as De Quervains disease (tendinitis in the long thumb abductor and the short thumb extensor) (Moore 1992).

3.2.3 Nerve entrapments

Nerve entrapments occur where a nerve may come under pressure or is exposed to mechanical friction (Lundborg 1988). Common nerve entrapment diagnoses in the working population are cervical nerve-root compression, 'neurogenic TOS' and carpal tunnel syndrome (CTS).

3.2.4 Degenerative joint disease and osteoarthritis

Degenerative changes in the cervical spine (spondylosis, i.e. spurs and/or disc degeneration) are common and 80% of the population may have degenerative changes at the age of 50 years on radiographs. Cervical spondylosis is related to job titles with high load on the cervical spine (Hagberg and Wegman 1987). However, the correlation between symptoms and cervical spondylosis is poor (Friedenberg and Mileer 1963; Lawrence 1969). In the shoulder glenohumeral joint osteoarthritis (Katevuo et al. 1985) and acromioclavicular joint osteoarthritis (Stenlund et al. 1992) have been associated to workplace factors.

All these neck and upper limb disorders can be caused by non-occupational factors, for example sports, leisure time activities or individual characteristics (malformations, genetic predisposing factors) (Figure 3.1). It is likely that all these disorders are caused by a combination of occupational, non-occupational and individual characteristics in the working population. Usually the epidemiological studies we have at hand, only consider one or two of these three likely routes to the disorder.

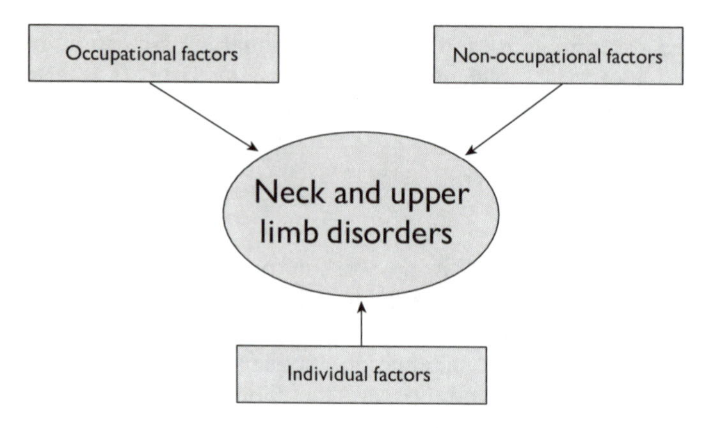

Figure 3.1 Neck and upper limb disorders may be attributable to occupational, non-occupational and individual factors.

The use of terms such as CTD (cumulative trauma disorders), OCD (occupational cervicobrachial disorders) and RSI (repetitive strain injuries) has been strongly criticized (Hadler 1989, 1990). Sometimes the terms have even been used as synonyms with disease terms such as CTS (carpal tunnel syndrome which is a compression of the median nerve at the wrist) or De Quervain's disease (an inflammation of the tendons to the long thumb abductor and the short thumb extensor at the wrist). This is not correct since both CTS and De Quervain's may be related to other factors than repetitive strain or to cumulative trauma. Also the use of CTD and RSI as synonyms for work related musculoskeletal disorders have been criticized since the term cumulative trauma disorders indicate a pathomechanism that usually is not proven. A musculoskeletal disorder may be caused by one trauma not necessarily a cumulative trauma. By using the terminology 'musculoskeletal disorders and work place factors', we recognize that these disorders are not specific to work and that workplace factors are one type of factor that are important in the causation of the disorder.

3.3 WHAT ARE THE EXPOSURES TO BE CONSIDERED?

Exposure in an occupational epidemiology context is usually referred to as conditions external to the workers, such as job demands and job requirements (Table 3.1).

Table 3.1 Definitions of exposure and related terms.

Term	*Definition*	*Reference*
Exposure	Proximity and/or contact with a source of a disease agent in such a manner that effective transmission of the agent or harmful effects of the agent may occur	(Last 1995)
Exposure	Any of a subject's attributes or any agent with which he or she may come into contact that may be relevant to his or her health	(Armstrong et al. 1992)
Exposure	Presence of a factor (substance) in the environment external to the worker	(Checkoway et al. 1989)
Cumulative exposure	Summation of intensity over time	(Checkoway et al. 1989)
Burden	Amount of a factor (substance) that exists in the body (organ) at a point in time	(Checkoway et al. 1989)
Dose	Amount of a factor (substance) that remains at the biological target at some specified time interval	(Checkoway et al. 1989)

Most epidemiological studies of neck and limb disorders have only a job-title as an indicator of the exposure. From the job-title the generic exposures factor are inferred. Generic physical risk factors for the neck and upper limb disorders are force, posture, repetition, contact stress and temperature. Each exposure factor can be assessed if the magnitude/intensity, duration and frequency are known. Whether repetition is a generic exposure factor is obscure, since repetition is a function of force over time, posture over time or contact stress over time. Other division into generic exposure factors related to neck and upper limb disorders have been material handling, vibration, and task invariability. Others have suggested the generic factor energy transfer as a single exposure factor and magnitude and duration of the factor determines the outcome (Hagberg et al. 1997). This has also been used as a way of distinguishing disorders from injuries, where injuries are more related to an accident process. Please note that the term accident denotes an event not a health outcome (Hagberg et al. 1997). Large amounts of energy during a short time will result in an injury where the latency time between the energy transfer and the outcome is usually shorter than 48 hours. Whereas low magnitude energy transfers during a long period of time this will result in a disorder and here the latency time will be longer than 48 hours. It is important that the epidemiological terminology being used in the ergonomic field is the same as in other fields. Thus it may be good to avoid terms such as 'internal exposures' and instead use the term burden or dose as expressions of the energy transferred in the individual.

3.4 WHAT ARE THE EXPOSURE RESPONSE RELATIONSHIPS FOR NECK AND UPPER LIMB DISORDERS?

In the last four years important reviews concerning the relation between workplace factors and neck and upper limb disorders have been published. 'Work related musculoskeletal disorders (WMSDS): a reference book for prevention' was published as a result of a consensus in an international group initiated by the Institute Recherche, Santé et Securité de Travail de Quebec (IRSST) (Hagberg et al. 1995). In 1995 conference proceedings from a conference sponsored by the American Academy of Orthopedic Surgeons was published 'Repetitive motion disorders of the upper extremity' (Gordon et al. 1995). Two years later, in 1997, NIOSH published 'Musculoskeletal disorders and workplace factors – a critical review of the epidemiological evidence for work related musculoskeletal disorders of the neck, upper extremity and low back.' This book is available on the Internet. The NIOSH publication reviewed and evaluated all epidemiological studies available published in peer review journals. The evidence was grouped into four categories: strong evidence, evidence, insufficient evidence or evidence of no effect. In the European Union an initiative was taken to evaluate whether there was enough evidence for causality between workplace factors and musculoskeletal disorders for the Union to encourage its member states to take actions. This document, 'Risk factors for work-related neck and upper limb musculoskeletal disorders' was published by the European Agency for Health at Work (Buckle 1999). The conclusion of the document is that there is enough evidence for action to be taken to improve workers' health.

3.5 EXPOSURE RESPONSE RELATIONSHIP ACCORDING TO A NIOSH DOCUMENT ON MUSCULOSKELETAL DISORDERS AND WORKPLACE FACTORS

In the NIOSH document 'strong evidence' and 'evidence' were found for different locations; especially the generic exposure factor posture was considered a risk factor for many symptom locations (Table 3.2). One of reasons for this is that posture is easier to study compared to repetition and force. Furthermore it is easier to infer postures from job titles than to infer other generic exposure factors.

Table 3.2 Approximate risk magnitudes of existing studies for generic risk factors to different musculoskeletal disorders (author's approximation median of the 3–5 publications with greatest risks). The risk magnitude express the ratio of the chance (odds) of having the disorder when exposed compared to not exposed. Data from NIOSH (Bernard 1997). NE = evaluated by NIOSH as not sufficient evidence or absence of evidence of work relatedness.

	Posture	*Force*	*Repetition*	*Vibration*
Neck non-specific pain	7	3	7	NE
Neck/shoulder non-specific pain	4	2	5	NE
Shoulder tendinitis	5	NE	3	NE
Elbow lateral epicondylitis	NE	6	NE	NE
Wrist/hand tendinitis	3	5	5	NE
Wrist/hand CTS	NE	4	4	(5)*
Wrist/hand				9*
Hand/arm vibration syndrome				

* Vibration exposure confounded by exposure to repetitive forceful grips it is uncertain whether vibration per se can cause CTS (author's opinion).

3.5.1 Neck and neck/shoulder

The outcome considered here is non-specific pain. Posture and repetition are indicated as the strongest risk factors. This is also consistent with hypothesis of pathogenesis of non-specific neck and neck/shoulder pain (Hagberg 1996).

3.5.2 Shoulder

For the outcome tendinitis posture was a strong risk factor. The magnitude in some studies was great. This is also consistent with the theory of the causation of the impingement syndrome (Neer 1994; Hagberg 1994). There are also other types of shoulder disorders that have been associated with work place factors such as osteoarthritis of the glenohumeral joint, osteoarthritis of the acromioclavicular joint and 'neurogenic TOS' (Katevuo et al. 1985; Stenlund et al. 1992; Toomingas et al. 1999, 1991).

3.5.3 Elbow

Here the outcome was lateral epicondylitis and force had the magnitude of approximately 6 and a prospective study showed an incidence ratio of approximately 7.

3.5.4 Hand/wrist

For the outcome of tendinitis posture, force and repetition had risk magnitudes of 3, 5 and 5 respectively. The outcome of carpal tunnel syndrome had a magnitude of 4 for repetition and for force and 5 for vibration. Please note that this risk magnitude is not adjusted for force or repetition, in a study by Silverstein et al. they found a risk of vibration of 5 but when adjusting for high force and repetition this risk dropped to 2 (Silverstein et al, 1987). The magnitude for the risk of hand-vibration syndrome was approximately 9 for vibration.

3.5.5 Other neck and upper limb diagnoses

In my clinical work I often have to evaluate patients with cervical nerve-root engagement or frozen shoulder. There is at present no consensus that epidemiological studies shows a relationship between workplace factors and these disorders. This could be due to the few studies that have been performed were poor or lacked the power to show any relationship. We have to remember that absence of evidence is not evidence of absence.

3.6 PSYCHOLOGICAL AND SOCIAL WORKPLACE FACTORS

In recent years psychological and social factors such as job strain (job demands in relation to job control) and supervisor support have gained attention. There is evidence that psychosocial factors through different mechanisms can cause musculoskeletal pain. Neck and neck/shoulder disorders were more related to psychosocial factors than 'peripheral' locations of disorders such as hand and wrist (Toomingas et al. 1997). It can be hard to separate psychosocial workload factors from physical factors by the workers (Josephson et al, 1996).

3.7 CONTROVERSIES CONCERNING NECK AND UPPER LIMB DISORDERS AND WORKPLACE FACTORS

In the last five years there has been some public debate concerning neck and upper limb disorders in relation to workplace factors. I have interpreted the debate as a miscommunication between epidemiologists and clinicians. From a clinical point of view you will see many patients where both the physician and patient are convinced that the disorder is caused by non-occupational activities or individual factors. Thus the clinician will object when you term a common disorder, e.g. non-specific neck/shoulder pain a work-related disorder. This could be due to a misunderstanding since non-specific neck/shoulder pain can be caused by occupational and by non-occupational factors. Furthermore the consensus of a critical review of the epidemiological evidence shows a substantial risk for these neck and upper limb disorders when there is exposure to certain risk factors. However, if the relative risk for a disorder is estimated to be 3, this implies that 67% of the disorder among the exposed is attributable to the exposure and that 33% of the disorder among the exposed is attributable to other factors.

In 1995 Vender and co-workers published a review stating that the epidemiological literature gave no support for work relatedness for upper extremity disorders (Vender et al. 1995). The basis for this was that at a literature search they found only 14 acceptable publications and those showed major flaws in 'sound medical diagnostic criteria'. In response to this article, an international group concluded that there were a number of good reviews that showed exposure-response relations in scientific literature to motivate preventive actions (Armstrong et al. 1996).

3.8 REFERENCES

Armstrong BK, White E, Saracci R. *Principles of Exposure Measurement in Epidemiology,* Monographs in epidemiology and biostatistics, vol. 21. Oxford University Press, Oxford, 1992.

Armstrong T, Buckle P, Fine L, Hagberg M, Sweeney MH, Kilbom Å, Laubli T, Punnet L, Sjogaard G, Viikari-Juntura E. Can some upper extremity disorders be defined as work-related. *J Hand Surg* 1996; 21A: pp 727–8.

Bernard BP, ed. *Musculoskeletal Disorders and Workplace Factors.* National Institute for Occupational Safety and Health, Cincinnati, 1997.

Buckle P. *Risk Factors for Work-related Neck and Upper Limb Musculoskeletal Disorders.* European Agency for Safety and Health at Work, www.osha.eu.int, 1999: 1–93.

Buchbinder R, Goel V, Bombardier C. Lack of concordance between the ICD-9 classification of soft tissue disorders of the neck and upper limb and chart review diagnosis: one steel mill's experience. *Am J Ind Med* 1996; 29(2): 171–82.

Checkoway H, PearceN, Crawford-Brown DJ. *Research Methods in Occupational Epidemiology.* Oxford University Press, New York, 1989: 344.

Friedenberg ZB, Miller WT. Degenerative disc disease of the cervical spine. *J Bone Joint Surg* 1963; 45-A: 1171–1178.

Gordon SL, Blair SJ, Fine L, eds. *Repetitive Motion Disorders of the Upper Extremity.* American Academy of Orthopedic Surgeons: Rosemont, 1995: 1–565.

Hadler NM, Work-related disorders of the upper extremity. Part I: cumulative trauma disorders – a critical review. *Occupational Problems in Medical Practice,* 1989; 4: 1–8.

Hadler NM. Cumulative trauma disorders an iatrogenic concept. *J Occup Med* 1990; 32: 38–41.

Hagberg M. Neck and shoulders disorders, in *Textbook of Occupational and Environmental Medicine,* L Rosenstock and MR Cullen, eds. WB Saunders Company, Philadelphia, 1994: 356–364.

Hagberg M. Exposure considerations when evaluating musculoskeletal diagnoses, in *Advances in Occupational Ergonomics and Safety,* A Mital et al. (eds). International Society for Occupational Ergonomics and Safety, Cincinnati, 1996: 411–15.

Hagberg M. *Neck and arm disorders. BMJ,* 1996; 313: 419–22.

Hagberg M, Wegman DH. Prevalence rates and odds ratios of shoulder neck diseases in different occupational groups. *Brit J Indust Med* 1987; 44: 602–610.

Hagberg M, Christiani D, Courtney TK, Halperin W, Leamon TB, Smith TJ. Conceptual and definitional issues in occupational injury epidemiology. *Am J Indust Med* 1997; 32: 106–116.

Hagberg M, Silverstein B, Wells R, Smith MJ, Hendricks HW, Caryon P, Pérusse M. *Work Related Musculoskeletal Disorders (WMSDs): a reference book for prevention*, I Kuorinka and L Forcier (eds). Taylor & Francis, London, 1995: 1–421.

Josephson M, Hagberg M, Wigaeus Hjelm E. Self-reported physical exertion in geriatric care. A risk indicator for low back symptoms? *Spine* 1996; 21: 2781–5.

Katevuo K, Aitasalo K, Lehtinen R, Pietila J. Skeletal changes in dentists and farmers in Finland. *Community Dent Oral Epidemiol* 1985; 13: 23–25.

Last JM, ed. *A Dictionary of Epidemiology*, 3rd edition. Oxford University Press, New York, 1995: 180.

Lawrence JS. Disc degeneration, its frequency and relationships to symptoms. *Ann Reum Dis* 1969; 28: 121–138.

Lundborg G. *Nerve Injury and Repair*. Churchill Livingstone, Edinburgh, 1988.

Moore JS. Function, structure, and responses of components of the muscle-tendon unit. *Occup Med* 1992; 7: 713–40.

Neer CS. Impingement lesions. *Clin Orthop* 1983; 173: 70–77.

Ranney D, Wells R, Moore A. Upper limb musculoskeletal disorders in highly repetitive industries: precise anatomical physical findings. *Ergonomics* 1995; 38: 1408–23.

Silverstein BA, Fine LJ, Armstrong TJ. Occupational factors and carpal tunnel syndrome. *Am J Indust Med* 1987; 11: 343–358.

Stenlund B, Goldie I, Hagberg M, Hogstedt C, Marions O. Radiographic osteoarthrosis in the acromioclavicular joint resulting from manual work or exposure to vibration. *Br J Ind Med* 1992; 49: 588–593.

Toomingas A, Nilsson T, Hagberg M, Lundström R. Abduction external rotation test among male industrial and office workers. *Am J Indust Med* 1999; 35: 32–42.

Toomingas A, Hagberg M, Jorulf J, Nilsson T, Burström L, Kihlberg S. Outcome of the abduction external rotation test among manual and office workers. *Am J Indust Med* 1991; 19: 215–27.

Toomingas A, Theorell T, Michélsen H, Nordemar R. Associations between self-rated psychosocial work conditions and musculoskeletal symptoms and signs. *Scand J Work Environ Health* 1997; 23: 130–139.

Vender MI, Kasdan ML, Truppa KL. Upper extremity disorders: a literature review to determine work-relatedness. *J Hand Surg* 1995; 20A: 534–41.

Viikari-Juntura E. Neck and upper limb disorders among slaughterhouse workers. *Scand J Work Environment Health* 1983; 9: 283–90.

Waris P, Kuorinka I, Kurppa K.e.a. Epidemiologic screening of occupational neck and upper limb disorders: Methods and criteria. *Scand J Work Environ Health* 1979; Suppl 3: 25–38.

Chapter 4

Psychosocial factors at work in relation to musculoskeletal conditions

Implications for job design and rehabilitation

Töres Theorell

4.1 INTRODUCTION

Psychosocial factors may comprise anything from personality to job organization.

The situation is further complicated by the fact that there are three different endpoints, one social (disability), one psychological (illness) and one medical/physiological (disorder). Although these are interrelated it creates confusion in the literature that they are often mixed. Similarly there are three different kinds of mechanisms which may tie psychosocial factors to symptoms in the musculoskeletal system (see below). To complicate matters even more, the mechanisms generating acute conditions in the locomotor system are to a great extent other than those perpetuating pain and creating chronic conditions. Often these two groups of mechanisms are mixed.

The three classes of mechanisms could be described in the following way:

(1) Physiological mechanisms which lead to ***organic changes***. An old theory that has been formulated in psychophysiology states that longlasting adverse life conditions enforce energy mobilization, *catabolism*, at the expense of restoring and rebuilding activities in the body, *anabolism*. The skeleton and the muscles are constantly being rebuilt to be adapted to the pattern of movements that the individual has. Muscle cells are being worn out and have to be replaced. Interestingly, several of the hormones that stimulate anabolism have their peak activity during the deep phases of sleeping. When sleep becomes more shallow, the activity of these hormones diminishes. This may show the importance of deep sleep to our possibility to protect our body from the adverse effects of longlasting energy mobilization.

One of the theoretical psychosocial models that have been studied in relation to psychophysiological activation and inhibition of anabolism is the demand-control model which states that high demands that high psychological demands and concomitant low decision latitude at work is a particularly adverse combination in relation to health effects of job conditions (Karasek and Theorell 1990).

Accordingly it has been shown in a longitudinal study that increasing job strain (high psychological demands in relation to possibility to issue control over the job situation) is associated with increasing sleep disturbance and gastrointestinal symptoms (Theorell et al. 1988). Sleep disturbance inhibits anabolism and gastrointestinal function is part of the anabolic function. In the same study increasing job strain was also associated with decreasing testosterone blood concentration which is even more directly associated with anabolism (Theorell et al 1990). In another study (Grossi et al. 1999) improving work situation for a group of police officers was associated with increasing blood concentration of testosterone. Thus, there is some evidence that increasing job strain is associated with decreasing anabolism. The evidence, however, must be regarded as incomplete. There are also studies which indicate that there is a direct association between job strain and energy mobilization. In the longitudinal study referred to above it was found that increasing job strain was associated with increasing blood pressure during activities at work (Theorell et al. 1988) and there is substantial evidence that job strain is associated with elevated blood pressure during working hours from other studies as well (Schnall et al. 1994). There are also studies which indicate that a low decision latitude is associated in itself with elevated catecholamine excretion in the urine, for instance a study by prison personnel (Härenstam and Theorell 1988).

Lack of social support – another aspect of the psychosocial work situation – has been shown to have sympathoadrenal correlates. The sympathoadrenal system consists of the adrenal medulla and the sympathic nervous system. These two stimulate the release of adrenaline and noradrenaline which help the body to mobilize energy. This is reflected in increasing concentrations of carbohydrate and lipid in the blood and also result in increased heart rate and blood pressure. Persons who report low social support at work have a high heart rate throughout day and night (Orth-Gomér and Undén 1991) and a decreasing social support at work has been shown to be associated with rising systolic blood pressure levels in a longitudinal study (Theorell 1990).

Longlasting inhibition of anabolism may adversely affect the vulnerability of several organ systems such as skeletal muscles, immune cells and gastrointestinal functions. This may certainly have significance to the development of acute organic conditions in the musculoskeletal system. If it is correct, a relatively small mechanical load that does not cause injury under normal conditions may do so during such a period. For instance a worker who has had adverse psychosocial conditions at work during several months may develop an injury at home when he rises from bed in the morning. Unfortunately this mechanism has been explored to a very limited extent in the literature.

Muscle tension that is reflected in electrical activity in muscles (EMG) is another important possible mechanism that may tie the psychosocial situation to risk of developing disorders in the locomotor system. Recent psychophysiological studies of monotonous work have indicated that the ergonomic loads typical of this kind of work elicit increased electrical muscle activity (EMG). However, when psychological loads are added, the electrical muscle activity increases considerably (Lundberg et al. 1993). This illustrates that psychological and physical demands may interact in complex ways in generating locomotor disorder.

Several indirect pathways may also be of importance to the relationship between psychosocial job factors and the development of work related behaviour. Such behaviour may be of marked importance to the development of acute musculoskeletal disorders.

Organic changes are also discussed in relation to chronic conditions in the musculoskeletal system. There is evidence that even the idiopathic or chronic pain syndromes with atypical localization of pain that constitute a major part of patients in rehabilitation may arise in well defined physiological processes, for instance spread of muscle tension which causes metabolic changes which may facilitate further spread of muscle tension etc. (see Johansson and Sjölander 1993).

(2) Physiological mechanisms that influence ***pain perception***. According to a rapidly growing literature, there is in chronic pain syndromes a strong element of psychological depression which influences the adrenocortical axis as well as hormonal systems such as the dopamin sensitive prolactin levels. The turnover of endorphins may also be also influenced. There is evidence from psychological literature that feelings of lack of control in the general life situation may increase pain sensitivity (see Maier et al. 1982; Feuerstein et al. 1987; Reesor and Craig 1987).

A recent study by our group has shown that pain threshold may be related to the demand-control-support model in a very complex way. Randomly selected men and women were asked to fill out questions about these dimensions in the description of their work. Their pain thresholds were measured on six different points in the neck and shoulders, first at rest before anything else happened in the laboratory, second in conjunction with an experimental stress test (the Stroop test) and finally after a new resting period. Our analyses indicated that as expected the pain threshold increased during acute stress, that those who reported a habitually high level of psychological demands at work had higher pain thresholds than others (i.e. were less pain sensitive than others) and finally that those who reported a low level of decision latitude in their habitual working situation had a lower pain threshold during experimental stress than others (Theorell 1993). The interpretation of these findings is very difficult. It seems plausible that subjects with high psychological demand levels who mobilize energy to be able to meet demands may also repress feelings of pain. This may increase risks of longterm development of pathological changes in locomotor tissues. On the other hand it is also plausible to assume that long-term exposure to low decision latitude may increase the risk of depression and that this could explain why subjects who describe their jobs in these terms are unable to raise their pain thresholds during acute stress situations.

(3) Sociopsychological conditions that are of significance to the individual's ***possibility to cope with the illness***. Such conditions are of central importance in the rehabilitation of subjects who suffer from pain in the locomotor system. Of particular interest for subject in working ages is illness behaviour in relation to sick leave. In the study by Theorell et al. (1991), psychosocial working conditions were studied in relation to outcome variables on three different levels, namely psychophysiological correlates (muscle tension, plasma cortisol, sleep disturbance), pain in the locomotor system and finally sick leave. Several interrelationships were found (after adjustment for gender, age, weight in relation to height and physical demands at work) both between the levels and between psychosocial factors and them. A statistically significant relationship was found between lack of possibility to influence decisions at work (authority over decisions) and sick leave – the lower possibility to influence decisions the more sick leave, and this relationship was not mediated by psychophysiological reactions or pain. Thus, some of the illness behaviour variables may actually be in the web of factors although they are independent of physiology and pain perception.

Recently the interaction between the individual coping pattern and demands at work has been explored as a potentially important factor in relation to musculoskeletal disorder (Eriksen 1998). It was found that a combination of high psychological demands and ineffective coping was associated with marked excess risk of having musculoskeletal disorder. It should be pointed out that coping (which is constituted by genetic factors and previous experiences) is moulded partly by the work environment. Accordingly the individual responds to stressors in the workplace and his/her experiences of the effects of these responses contribute to a change in coping in the same kinds of situations in the future (see for instance Theorell et al. 1999).

4.2 OUTSIDE WORK FACTORS

For middle aged men, psychosocial factors outside work, for instance when a spouse becomes ill (Theorell et al. 1975) may be of importance to organic changes. Worries about the spouse lead to increased energy mobilization and catabolism and decreased anabolism as well as increased muscular tension. Movements in awkward positions (which become more common in this situation since the healthy spouse has to do work at home that he has not been doing to any great extent before!) may create iterated trauma to and vulnerability in the musculoskeletal system. This may increase the likelihood that injury may arise at work. It may also affect pain perception (depression may lower pain thresholds) and illness behaviour (there is increased need to stay home to help the spouse).

Outside work factors may be of particular importance to the female working force. In a recent study Josephson (1999) showed that home work in itself may be of substantial importance to the risk of seeking care for acute lumbar and/or neck/shoulder pain. Women with children below age 7 who estimated that they had at least 40 hours a week of unpaid work at home had a relative risk of 6.2 (95% confidence limits 1.6–24.0) compared to women with no children below age 7 estimating their hours spent in unpaid work to be below 40 hours per week.

4.3 PHYSICAL LOAD AND PSYCHOSOCIAL FACTORS

One of the reasons for the inconsistent findings in different studies may be that groups with markedly different amounts and kinds of ergonomic and psychosocial stressors have been studied. Efforts have sometimes been made to adjust statistically for ergonomic load in the study of locomotor symptoms in relation to psychosocial stressors (see for instance Theorell et al. 1991). If strong interactions take place between ergonomic and psychosocial stressors adjustment is of no help in the analysis. An example from an ongoing study is the following:

A study of the determinants of frequency of pain in shoulders and low back, respectively, in a sample of female health care personnel was based primarily upon a questionnaire (Ahlberg-Hultén et al. 1994). Information regarding demands, decision latitude, support as well as social relationships between work mates and with superiors was included as well as estimations of ergonomic load in the various wards (outpatient care of paediatric children, infectious ward and emergency ward) and marital state and number of children. Multivariate analyses (multiple logistic regression) showed that low authority over decisions was a statistically significant determinant of low back pain, regardless of other determinants – the lower authority the employee felt that she had the higher likelihood that she would report low back pain. Similarly, a measure of social support at work was the only statistically significant determinant of pain in the shoulders – the poorer support at work the greater likelihood that an employee would report pain in her shoulders. In univariate analyses the group (after trichotomy of the study group) of women with the worst authority over decisions had 3.0 times higher likelihood of reporting low back pain and the group of women with the worst social support (also trichotomy) had 4.4 times higher likelihood of reporting shoulder pain than other women. There was only marginal overlap between the groups, the relationship between reporting low back pain and shoulder pain was not statistically significant and the contingency coefficient (roughly corresponding to correlation) was trivial (=0.15). Accordingly we are dealing for the most part with two different groups of subjects. A reasonable speculation is that emotionally loaded situations often arise when ill patients are moved from supine to sitting positions or from one bed to another. These situations require skilful cooperation between patient and personnel but also between staff members. If the social atmosphere in the ward is poor, these situations may become more unpredictable and the likelihood of sudden unexpected physical load during lifting increases. This may be more likely to occur in the neck/shoulder region when forward bending positions and use of arms and shoulders are common. Low back pain may arise in other kinds of situations with heavy lifting of objects in which the social atmosphere may be less important but in which feelings of frustration created by poor possibilities to influence decisions may play a role. Job strain has been shown to be closely associated with heavy physical work in women but not in men (Josephson 1998). Accordingly, women with 'job strain' were much more likely than other women (DR=4.1; 1.6–11.0) to have heavy physical workload.

4.4 ADDITIONAL REASONS FOR CONTROVERSIES IN THE PSYCHOSOCIAL FIELD

Apart from the fact that three different mechanisms are closely involved in one another there are also other reasons for diverging findings in the scientific studies of relationships between psychosocial factors at work and musculoskeletal disorders:

(1) Measurements need to be refined since some of the studies have been based upon poor assessments. There is considerable activity internationally in the improvement of psychosocial including observation techniques, interview guides and questionnaires.
(2) The time axis in the illness processes has not been given adequate attention. There are considerable differences between subjects who have had pain for a long time and those who have had a recent onset of musculoskeletal pain.
(3) Gender and age effects have been studied in an insensitive way. For instance, age and gender have often been used as confounders. A more adequate analytical strategy is to analyse men and women separately and also to separate younger and older subjects in the analyses.

(4) Biologically it may not be meaningful to try to separate the effects of psychosocial factors from those of physical factors since the brain makes no such distinction. In future studies it will be important to study joint effects of psychosocial and physical factors.

4.5 EMPIRICAL ILLUSTRATIONS FROM RECENT STUDIES

Some of the results from a recent study, the 'MUSIC-Norrtälje' (Musculoskeletal Intervention Center in Norrtälje which is a rural/suburban part of the greater Stockholm area) are of interest as illustrations since this study has included physical and psychosocial factors in 'new' cases of lumbar pain and neck/shoulder pain as well as in a comparable population sample – referents – without such pain. In a subsample of cases and referents, endocrine and immunological factors were studied as well.

The MUSIC-Norrtälje was initiated by a research group at the National Institute for Working Life Research with participation from the Institute for Environmental Medicine and the Department of Occupational Health at the Karolinska Institute as well as the National Institute of Psychosocial Factors and Health. The study was designed to answer questions regarding the etiology of lumbar pain and neck/shoulder pain. The assessments were carefully selected after validity and reliability tests.

The cases were recruited according to a special case finding method. All potential caregivers, including non-traditional ones such as chiropractors and massage experts, were contacted and asked to contribute with information about all 'new cases' that consulted them. A 'new case' was operationally defined as a subject who consulted a caregiver for an episode of lumbar pain or neck/shoulder pain without having done so during a period of six months preceding the consultation. The 'case' was then contacted by the MUSIC-Norrtälje team who spent half a day with him/her doing interviews and examinations as well as asking him/her to fill out questionnaires regarding physical loads and psychosocial conditions at work and outside work. At the same time at least one referent living in the same region who was comparable with regard to age and gender was selected from the population register. The referent was contacted and asked to go through the same procedure as the case.

The participation rate in the case group is not known. In the referent group the participation rate among women was 80% and among men 78%. There were 311 male and 375 female lumbar pain cases and 118 male and 271 female neck/shoulder pain cases who completed all the different parts of the study. The number of male referents varied between 589 and 662 and the number of female referents between 796 and 849 depending on which comparison that was made.

In multivariate analysis (Vingård et al. 2000) the most important predictors of being a male low back pain case were two ergonomic, 'forward bending' >60 min/day (RR = 1.8; 95% confidence limits 1.1–3.1) and 'heavy lifting' (RR = 1.4; 1.0–2.0), and two psychosocial factors, 'poor job satisfaction' (RR = 1.8; 1.0–3.1) and 'no possibility to learn and develop and mostly routine work' (RR = 1.8; 1.0–3.1). For females, only the ergonomic factors remained as important predictors of lumbar pain in the multivariate analysis, 'metabolic demand' MET >3 (RR = 2.0; 1.2–3.2) being the most important one. Thus, in male lumbar pain cases, both physical and psychosocial factors showed independent predictive value, whereas in female lumbar pain cases only the ergonomic factors remained as independent predictors. A different result emerges when combinations of physical and psychosocial conditions were examined. Among women the combinations 'high physical load and job strain' (after adjustments for age and previous low back pain episodes lasting for as least 3 months) as well as 'high physical load and low decision latitude' were all associated with markedly elevated risk of being a lumbar pain case (RR = 3.6; 1.3–10.9, RR = 2.2; 1.0–4.9 and RR = 2.8; 1.4–6.0). For men these combinations were not associated with elevated risks.

The findings regarding neck/shoulder pain cannot be reported in detail yet (Wigéus-Hjelm et al. 1999). However, in multivariate analysis both for men and women physical load factors as well as psychosocial conditions are independent predictors of neck/shoulder pain. Both the physical load and the psychosocial factors differ from those predicting low back pain. As with low back pain, the patterns differ markedly between men and women.

The design of the MUSIC-Norrtälje makes it possible to calculate the 'attributable portion' associated with a risk factor that has been established in multivariate analysis. The attributable portion is the prevalence of the risk factors multiplied by the ratio between the excess risk and the relative risk. The more common the risk factor and the higher the relative risk, the higher the attributable portion. The attributable portion corresponds to the proportion of cases that could be prevented if the risk factor would be eliminated. Some examples from the MUSIC-Norrtälje:

The attributable portion in male lumbar pain cases associated with work in a forward bent position for at least 60 minutes a day was 13%. For women the strongest lumbar pain factor, MET >3, corresponded to 9%. The pronounced excess risk (RR = 3.6) associated with the combination of high physical load and job strain in women, corresponded to an attributable portion of 8%. However, for several psychosocial factors the attributable portion is considerable. This is particularly true in the neck/shoulder cases. For instance, in men the attributable portion of a poor social support at work associated with being a neck/shoulder case was 14%. For 'low demands in relation to competence' the corresponding attributable portion for being a male neck/shoulder case was 11%.

The epidemiological findings in the MUSIC-Norrtälje study give support to the hypothesis that psychosocial factors and thus long-lasting arousal and/or inhibition of anabolism may contribute to the etiology of acute pain. In male lumbar cases the combination of physical load and job strain was important. In neck/shoulder cases both psychosocial and ergonomic factors may be important but the factors are gender specific and only partly confirm the classical demand-control-support model (Wigeaus-Hjelm et al. 1999).

4.5.1 Physiological assessments

In a subsample in the MUSIC-Norrtälje venous blood samples were drawn in the morning of the MUSIC-examination and at noon. Blood samples were analysed with regard to the blood concentration of

1. ***Cortisol*** which reflects energy mobilization. In acute 'arousal' situations, the concentration rises in healthy subjects. In subjects with 'chronic fatigue syndrome', on the other hand, low cortisol levels with small circadian variation are found. The normal pattern in healthy subjects is a high morning level and a decreasing concentration during the day. Low morning levels with small circadian variations as in chronic fatigue syndrome may be a consequence of a long-lasting adverse life situation – 'chronic stress'.
2. ***Interleukin-6*** (IL-6) which is part of the communication between the immune system and the brain. The interleukines react both to infection/inflammation and to psychological stress.
3. ***Beta-endorphin***, one of the body's own analgesic compounds.
4. ***DHEA-s*** which reflects anabolic activity.
5. ***MHPG*** which reflects the activity in the sympatho-adreno-medullary system (adrenalin and noradrenalin) which has an important function in physiological arousal.

All these substances are discussed in (Hasselhorn et al. 1999).

The following significant results were found (Theorell et al. 2000; Hasselhorn et al. 1999).

Relationships between psychosocial conditions and endocrine/immunological factors after adjustment for a number of factors (Hasselhorn et al. 1999; Theorell et al. 2000) including age:

In men a low decision latitude at work was associated with a high midday IL-6 concentration. ***In women*** first a low level of job satisfaction was associated with a high IL-6 concentration, second a high level of job strain with high IgE-concentration, third low support at work with high DHEA-s and finally negative life events during the past year with a small difference between morning and midday cortisol. Thus, particularly in women there was evidence of an association between long-lasting adverse life conditions on one hand and 'exhausted' HPA axis, and an activated immune system on the other hand. One association, low support–high DHEA-s, was, however, in the 'wrong direction'. In men we only found support for the relationship between low decision latitude and an activated immune system.

Both in men and women there was a strong relationship between a small difference between morning and midday cortisol and a high IL-6 concentration. Thus an 'exhausted HPA' axis was associated with an activation of this part of the immune system.

Differences between cases and referents: Women with low back pain had a smaller difference between morning and midday cortisol and female cases (both low back pain and neck/shoulder pain) had higher IL-6 concentrations than female referents (Theorell et al. 2000).

4.5.2 Physiological factors associated with persisting pain and disability

Female low back pain cases with persisting high disability scores (up to 6 months after the MUSIC examination) had at the time of the MUSIC examination a low average concentration of DHEA-s, beta-endorphin and MHPG. Female neck/shoulder pain cases with a persisting disability had a high average concentration of IL-6. Male low back pain cases with persisting disability had high IL-6 concentration (Hasselhorn et al. 1999).

Accordingly there was evidence particularly in female cases with persisting disability of low anabolic activity, low pain regulation and low adrenomedullary as well as high interleukin-6 activity in the immune system. Again the findings were weaker in men than in women.

These observations on psychophysiological relationships are based upon a cross-sectional study with all its weaknesses. It cannot be established whether the changes in physiological parameters precede or follow the illness. In addition, the samples were relatively small (cases and referents together, 141 women, 102 men). The fact that the follow-up indicated significant relationships between 'stress physiology' – for instance low anabolic activity – and persisting disability adds a valuable prospective component to it. The results have to be replicated in other studies.

4.6 JOB INTERVENTIONS

If there are relationships between psychosocial conditions at work and musculoskeletal disorders it would be beneficial to improve the work organization in the direction of elevated decision latitude and improved social support. Psychological demands may be harder to influence in an increasingly high competition, but it is necessary also to monitor demands and intervene when psychological demands reach excessively high levels. Of particular importance to job intervention research in general is that decreased decision latitude may be associated with increased risk of development of myocardial infarction in middle aged men within a five year period (Theorell et al. 1998). Thus the discussion is not confined to musculoskeletal disorders only. Health may be positively affected in several ways by improved work organization.

Quality of working life experiments have been going on for at least two decades. They have been evaluated in productivity - but not frequently in health terms – and it has been shown that they have very often been associated with a lasting increase in productivity (see for instance Kopelman 1985). They have been building on principles outlined above. Increase in intellectual discretion has been achieved by means of job enrichment or increased worker responsibility for the complete product. Increase in decision authority has been achieved by means of job enlargement or flattened organization hierarchy. Increase in social support, finally, has been achieved by means of improved feedback and formation of more cohesive work groups. During later years, both in North America and in Scandinavia, the starting of regular staff meetings for systematic discussion of important decisions regarding work routines and goals has been used as a tool for increased decision latitude and social support.

The occupational health care team has often been an important vehicle in Scandinavian job redesign (see for instance Wallin and Wright 1986). The occupational health care team has the possibility to work both with the structure of the company and the individual workers. It has the possibility to carry out individual measurements of psychological and somatic health indicators and to monitor these through the change process. It has been the experience of health oriented job redesigners that the concomitant emphasis on structure and individuals is helpful both for individual motivation to follow health promotion advice (which is strengthened when workers discover that there is willingness in management to improve the structure) and for success in instituting structural change (which will be benefited if individuals are strengthened).

The redesign process will now be illustrated by a case study. The experiences of the case will be discussed and put into a larger framwork.

The job redesign process was prompted by current demands for effectiveness and speed in public service, but also by concerns caused by a high sick leave prevalence among the employees. Postal processing has been a public service given by the state in Sweden. No competition has existed so far, but recently small-scale efforts to create competing postal service organizations have started and due to national financial problems, discussions regarding the effectiveness in the governmental mail system have taken place.

In the Swedish postal system the occupational health care teams have been quite active in the psychosocial field (Previa, previously Statshälsan). It has developed both a system for exploring the psychosocial work environment and a system for instituting beneficial organizational changes in it. These programmes that have been designed for state employees in general have been extensively used in the country and have also been used by other organizations.

This organizational change was initiated by the regional head office in Spånga/Kista. The intention was to decrease the prevalence of musculoskeletal disorder which was common among employees.

Postal delivery is a physically demanding job. The relative physical demand increases with increasing age. Accordingly it is uncommon to work as a postman or postwoman after the age of 50. The traditional way of reducing physical demand has been to reduce ergonomical loads. There is also a possibility to change the work organization, however, since increased variation may reduce tissue strain and improved psychosocial climate may decrease vulnerability (see above). The health consequences of such changes in work organization, however, have not been studied extensively. Increasing observation of dynamic and less stable organisations in which several changes take place concomitantly has been recommended from this point of view (see Silverstein 1992). If many such evaluations are published a total scientific evaluation will be facilitated. In line with this recommendation, the evaluation of this experiment in postal delivery was performed (Wahlstedt et al. 1996).

The intervention was based upon the following practical premises: the postal station in Spång/Kista close to Stockholm started in 1975. Since this region has been expanding, the numbers of inhabitants, companies and postal districts have increased. In 1988, the planning of a division of the station started. From the summer in 1990 the plans became more concrete since convenient localities became available. The two new stations started in April 1991. Employees were given the option of choosing one of the two stations. Accordingly, the more traditional organization was planned in Spånga and the more profound one in Kista. There were changes in both places, however. Table 4.1 summarizes the changes in them.

Table 4.1 Changes in Spånga (traditional) and Kista (modern).

	Spånga	*Kista*
New localities	0	++
Restoration of localities	+	++
New sorter tables	+	++
Separate rooms for the teams	0	++
Organizationally cohesive teams	+	++
Increased foreman responsibility	+	++
Budget responsibility to foreman	++	++
Recruitment of temporarily employed to foreman	++	++
Planning of vacations to foreman	0	++
Bonus system	0	++

0 = unchanged, + = little change, ++ = drastic change.

The most important difference between the two work sites was that the foremen in the modern station had greater responsibility than those in the traditional one. Another important difference was that the employees in the modern station had joint responsibility for all the districts of the team whereas they had responsibility only for their own district in the traditional station. In the modern station the teams also had joint responsibility for partially new work tasks such as delivery of large collections of letters, sorting of boxes as well as handling of precious post and parcels. Two persons who should be available as buffering resources when short leaves occurred were appointed in the modern station. They were stationed in one of the teams but should be available for the other teams in cases of emergency as well. Time was allocated for planning and foreman activities in the new station, and cohesiveness was strengthened in this station by means of a team based bonus system. Each team in the new station had a room with windows for sorting.

All delivery men and women in the new station used bicycles when delivering post, whereas some of the employees in the traditional station used cars and some used bicycles – as previously – also after the changes had been instituted.

The employees had influence over the intervention process particularly from one point of view: since the regional office reasoned that the changes instituted in one of the sites were profound (and according to the hypothesis more effective and therefore important to test) and could be perceived as threatening, the employees were given the opportunity to choose one alternative.

It was possible to perform a before/after comparison both in the intervention group and in a group with less extensive intervention. The participants in the intervention group moved to a new station and went through a change in work organization. Employees were allowed to select which station they would belong to. Measurements were performed before intervention and twelve months later (1991 and 1992 respectively).

The main question in the evaluation of the mail delivery intervention was: Can improved work organization result in decreased prevalence of locomotor disorder? The study group consisted of 106 persons after exclusion of those who had served as assistants working only every second Saturday, those who were on long-term sick leave and those who had stopped working during the study period. Eighteen persons did not answer the first questionnaire in 1991 and six did not answer the 12-month follow-up questionnaire in 1992. Accordingly the final study group consisted of 82 persons. The persons who chose to move to the modern station (n = 27) were comparable with regard to gender and age to those who stayed in the traditional station (n = 55). Those who moved had in general been working in postal delivery for a significantly shorter period than those who stayed (median 2 and 4 years respectively).

A Swedish established self-administered questionnaire for the recording of psychological demands (five questions), intellectual discretion (four questions) and authority over decisions (two questions) (Theorell et al. 1988) was used. Cronbach's alpha coefficient for psychological demands and decision latitude was 0.84 and 0.83 respectively. In addition, on the basis of factor analysis, two indices were constructed for the measurement of contact with superiors (five questions) and contact with team mates (four questions). The Cronbach alpha coefficients were 0.78 and 0.67 respectively.

Symptoms from the locomotor system were recorded by means of a self-administered questionnaire (the Nordic Council questionnaire, see Andersson et al. 1984) which has been well established in previous research.

Ergonomic conditions (repetitive arm movements, uncomfortable standing, uncomfortable sitting, leaning and twisted postures, heavy lifting, repetitive leg work, uncomfortable arm work) did not change significantly according to the questionnaires.

Social support increased significantly in the modern station only and psychological demands decreased significantly in both stations. Authority over decisions increased in the traditional but not in the modern station – somewhat contrary to expectations. A combined index of neck, shoulder and chest back pain as well as shoulder and chest back pain decreased significantly in the modern (by 25%) but not in the traditional station (where the non-significant decrease amounted to 12%). The difference in improved prevalence amounted to 13% to the advantage of the new station (although such an assessment is of course rather imprecise). The favourable musculoskeletal changes were more pronounced in participants below age 35 than among those above this age.

4.6.1 Comments

The evaluation was not a classical one in the sense that two randomly selected samples were followed and compared with regard to outcome after the intervention. Two groups were compared. These two groups were self-selected and there is a possibility that more healthy and satisfied workers could have selected one of the two alternative stations. The available information speaks against this interpretation, however. Management initiated the changes. External consultation was used only for the evaluation. Accordingly the experiment reflects a real life situation which could be translated into viable follow-up experiments in other sites.

The theoretical background of the instituted changes was the demand-control-support model (Karasek and Theorell 1990). There was emphasis on improvement of authority over decisions. In the postal station the foremen were given increased authority in both stations but particularly so in the modern station. Since the foremen were closer to the individual delivery workers, this could have resulted in improved authority over decisions for the individual worker. In both cases the organizational changes could also result in improved intellectual discretion.

The other main emphasis was on improved social support/climate. Improved social support is often difficult to disentangle from increased authority over decisions. Frequently, an improved sense of individual control is obtained by means of collective actions of the total working group. In the modern postal station the corresponding tool was the formation of working groups with joint responsibility for several districts, allocation of these groups to separate rooms and financial group bonus.

Postal delivery is a highly variable activity that is physically demanding and takes place in direct interaction with customers. This may to some extent explain why the organizational changes and their results were more clear with regard to social support and interaction than for decision latitude. In several previous experiments, effects have been documented after group feedback and group intervention on either different aspects of authority over decisions/intellectual discretion (Gardell and Svensson 1981; Karasek and Theorell 1990, p. 211) or on social support (Orth-Gomér et al. 1994; Theorell et al. 1995). In previously published experiments, decreased (Orth-Gomér 1994; Karasek and Theorell 1990, p. 211) as well as increased psychological demands (Gardell and Svensson 1981) have been reported following the changes.

The main health outcome result in the delivery stations was that neck/shoulder/chest back pain decreased. This is consistent with observations that indicate that good social support at work may have a statistically significant protective effect in relation to locomotor disorder (Ahlberg-Hultén et al. 1995; Bongers et al. 1993).

Even in Sweden with its long tradition of work redesign, very few studies have actually documented health consequences of this type of job intervention. Most of the published studies have been based upon self-administered questionnaires. One study of cardiovascular risk factors (Orth-Gomér et al. 1994; Theorell et al. 1995) showed that a similar program (which was evaluated in a controlled study) was followed by a reduced ratio of LDL/HDL cholesterol in the experimental but not in the comparison group. As in all experiments of this kind, the design was not 'clean' in the sense that all other changes had been ruled out, but the findings could be interpreted to mean that job redesign aiming at improved social relationships and decision latitude could lead to decreased cardiovascular risk. In that study, it was also shown that the psychosocial job changes were more beneficial in worksites that had an active attitude to the change. In those with a more passive attitude, feedback from superiors deteriorated and the psychoendocrinological arousal – as reflected in serum cortisol levels during the intervention – was more pronounced.

Few studies have been published regarding postal processing and its psychosocial work environment. One of the largest studies published in this area was the one by Amick (1991). On the basis his study of nearly 5000 postal sorters, Amick suggested that influencing supervisor support is the most effective way to affect a person's job satisfaction and level of psychosomatic symptomatology.

4.6.2 Important points

(1) ***Job changes require considerable time.*** Already when an exploration (which is always regarded by the employees as a start of a change process) is planned, the team must have resources and allocate time to follow-up of the consequences. Social changes always take more time than expected! Months and years rather than days and weeks are required.
(2) ***Different groups of actors have to be informed*** well in advance before and continuously during the process. Processes that arise as needs formulated by the workers are always more likely to be successful than top-down processes! (see for instance Gustavsen and Hunnius 1981). The managers and superiors have to be informed regarding important findings and developments during the process, for instance when group feedback is planned.
(3) ***Group feedback and discussion are important components*** in the change process. The importance of group feedback is illustrated by the findings in a study of six occupational groups, with each participant being monitored on four occasions during a year. Very small changes in the perceived work environment were reported in relation to individual feedback occasions. After group feedback had taken place, on the other hand, measurements indicated that psychological demands had decreased and intellectual discretion increased significantly. More conflicts and more awareness of work environment problems were also reported after the group feedback occasion (Karasek and Theorell 1990).

Group feedback and discussion has to be organized in a systematic way. What is found in the exploration is compared with other groups which are used as reference. Both strengths and weaknesses are identified. The positive aspects are important since the workers need pride and self esteem. But the worksite has to identify problems that are important to ameliorate. In the ideal situation, special employees are elected who have the task (within a specified time frame) of formulating practical solutions. These solutions are discussed on the next occasion. Several such groups may work concomitantly on different specified problems.

(4) ***Mental preparation for conflicting opinions about solutions.*** After the initial phases of 'engagement' and 'search', the 'change' phase occurs. During this phase there are always different opinions about solutions. If

the initiators of the process are unprepared for this diversity of opinions (and potential serious conflicts!) they may not facilitate the redesign in a helpful way.

4.6.3 Conclusion on job redesign

This review and case description illustrate that work redesign for the improvement of the psychosocial environment has had a strong position in Sweden's working life for a long time. Swedes have gathered experiences and know-how. Due to a recent societal financial crisis, however, health promotion work in relation to work organization has had a weaker position during later years. The case description illustrates that it is possible to improve the work organization of public service and that such changes may benefit employee health.

Table 4.2 Self-reported psychosocial/organizational conditions at the old (A) and the new (B) station 1991 and 1992 (n = 82).

Index (possible variation)	*Stayed in A n = 55*		*Moved to B n = 27*		*Total group n = 82*	
	1991	*1992*	*1991*	*1992*	*1991*	*1992*
Social support[a] (16–64)	29.1	27.3	28.2	25.8[1]	28.8	26.8[2]
Contact with supervisors[a] (5–20)	7.7	7.7	6.8	7.4	7.4	7.6
Contact with work mates[a] (4–16)	6.2	6.5	6.2	5.7	6.2	6.2
Psychological demands[b] (5–20)	12.9	11.6[3]	13.2	11.6[4]	13.0	11.6[5]
Intellectual discretion[b] (4–16)	10.0	9.8	9.2	9.9	9.8	9.9
Authority over decisions[b] (2–8)	5.3	5.8[6]	5.8	5.7	5.5	5.8

[1] $p = 0.04$ [2] $p = 0.02$ [3] $p = 0.001$
[4] $p = 0.006$ [5] $p = 0.000$ [6] $p = 0.03$.
[a] The higher the score the lower degree of support.
[b] The higher score indicates more demands. more intellectual discretion and more authority over decisions.

4.7 PSYCHOSOCIAL FACTORS IN REHABILITATION

No comprehensive review of psychosocial factors in relation to rehabilitation will be given in this chapter. It should be pointed out, however, that there is evidence that psychosocial working conditions may be particularly important in relation to long-term prognosis after the onset of spine pain – and perhaps more important than they are in relation to acute onset. Leino and Hanninen (1995) have published a ten-year follow-up of 900 white- and blue-collar workers. They showed that changes in morbidity (pain and disability) related to musculoskeletal disorder were predicted better by psychosocial job conditions (possibility to control working conditions for

Table 4.3 Prevalence (%) of symptoms from the local motor system at the old (A) and the new (B) station 1991 and 1992 (n = 82).

Anatomical region	*Stayed in A* n = 55		*Moved to B* n = 27		*Total group* n = 82	
	1991	*1992*	*1991*	*1992*	*1991*	*1992*
Neck	40	26	46	44	42	32
Shoulders	51	37	52	28[1]	51	34[2]
Thoracic spine	29	21	33	8[3]	30	17[4]
Lumbar spine	43	30	42	36	43	32
Arm[a]	31	24	27	24	30	24
Leg[b]	48	54	41	29	46	46
Neck/shoulders/thoracic spine[c]	56	44	69	44[5]	61	44[6]

[1] $p = 0.02$ [2] $p = 0.009$ [3] $p = 0.02$
[4] $p = 0.02$ [5] $p = 0.04$ [6] $p = 0.006$.
[a] Symptoms from elbows and/or symptoms from wrists/hands.
[b] Symptoms from hips and/or symptoms from knees and/or symptoms from ankles/feet.
[c] Symptoms from neck and/or symptoms from shoulders and/or symptoms from thoracic spine.

the employees as well as social relations) than by other factors. Ekberg and Wildhagen (1996) followed 93 subjects who had recently developed neck and shoulder pain for one year. Multiple regression analysis indicated that long-term sickness absence was largely associated with work conditions rather than with individual characteristics. Therefore, the results underscore the importance not only of treating the individual with musculoskeletal disorders, but in particular of improving his or her work conditions. There is no unanimous agreement about this in the literature, however. Estlander et al. (1998), for instance, did not find that the psychosocial variables in their two-year follow-up study of 452 working subjects were independently predictive (in multivariate analysis) of development of and changes in musculoskeletal pain. They concluded that the present research paradigm is not sufficient for the study of these complex relationships. Thus more research is needed.

Rehabilitation programmes that have included worksite orientation as well as individual rehabilitation have usually been more successful than other programmes. Linton and Bradley (1992) for instance reported an 18-month follow-up of a secondary prevention programme for back pain of recent onset. Results showed that subjects had significantly less pain, used fewer medications, and were more active at 18-month follow-up than at baseline. All subjects had returned to work, and one-third had no pain-related work absences during the follow-up. A cost-benefit analysis indicated substantial economic savings when follow-up sick-listing data were compared with estimates based on an increasing trend for pain-related absenteeism found during the baseline period. Hindrance factors reported by subjects were related to personal time-management and workplace factors, especially psychosocial aspects of the work environment. It was concluded that the secondary prevention programme was effective and that future maintenance programmes should focus more on personal time-management and workplace factors.

The main conclusion regarding rehabilitation in relation to psychosocial factors is that insufficient attention has been given to this in practice and research and that there are indications in the literature pointing at the importance of them. It is obvious that a person who is strengthened both physically and psychologically during a rehabilitation process will have difficulties and may have an unfavourable prognosis if he or she returns to a work environment with poor social relationships and poor possibilities to influence the working situation.

4.8 CONCLUSIONS

The following general conclusions were made:

(1) Different theoretical models have to be used in the study of etiology of musculoskeletal disorders versus long-term outcome of these disorders
(2) The global concepts psychological demands, decision latitude and social support at work seem to be useful, and there is empirical support for the assumption that these dimensions are of relevance to the development of musculoskeletal disorders and symptoms. Relatively simple theoretical models are preferable in preventive practical work.
(3) There is a complex interaction between physical and psychosocial job conditions.
(4) Different parts of the demand-control-support model are of importance in different populations and possibly also to different diagnostic categories within the musculoskeletal system.
(5) More physiological studies are needed in the exploration of possible pathways.
(6) Both cross-sectional and longitudinal studies will be needed – recall bias is a problem in cross-sectional studies and lack of relevant information for the most recent period preceding the onset of locomotor disorders or symptoms is a problem in longitudinal studies.
(7) Precision in the psychosocial exposure measurements needs to be developed preferably by means of direct comparisons between observations, interviews and questionnaires.

4.9 REFERENCES

Ahlberg-Hultén G, Sigala F, Theorell T. Social support, job strain and pain in the locomotor system among female health care personnel. *Scand J Work Env Health* 1995; 21: 435–439.

Amick B. Structural determinants of the psychosocial environment: introducing technology in the work stress framework. *Ergonomics* 1991; 34: 625–646.

Andersson G, Biering-Sorensen F, Hermansen L, Jonsson B, Jorgensen K, Kilbom Å, Kuorinka I, Vinterberg H. *Nordiska frågeformuläret för kartläggning av yrkesrelaterade muskuloskeletala besvär* [The Nordic questionnaire for the assessment of work related musculoskeletal symptoms]. *Nordisk Medicin* 1984; 99: 54–55.

Bongers PM, de Winter CR. *Psychosocial Factors and Musculoskeletal Disease*, mimeograph, Nederlands Instituut voor Praeventieve Gezondheidszorg TNO, 1992.

Bongers PM, de Winter CR, Kompier MA, Hildebrandt VH. Psychosocial factors at work and musculoskeletal disease. *Scand J Work Env Health* 1991; 19: 297–312.

Ekberg K, Wildhagen I. Long-term sickness absence due to musculoskeletal disorders: the necessary intervention of work conditions. *Scand J Rehabilitation Medicine*, 1996; 28(1): 39–47.

Eriksen HR. Stress and Coping: Does it really matter for subjective health complaints? Thesis. University of Bergen, Norway, 1998.

Estlander AM, Takala EP, Viikari-Juntura E. Do psychological factors predict changes in musculoskeletal pain? A prospective, two-year follow-up study of a working population. *J Occupational Environmental Medicine* 1998; 40(5): 445–53.

Feuerstein M, Papciak AS, Hoon PE. Biobehavioral mechanisms of chronic low back pain. *Clin Psycho Rev* 1987; 7: 243–273.

Gardell B, Svensson L. *Medbestämmande och självstyre: En lokal facklig strategi för demokratisering av arbetsplatsen* [Worker participation and autonomy: A local union strategy for democratization of the workplace]. Prisma, Stockholm, 1981.

Grossi G, Theorell T, Jürisoo M, Setterlind S. Psychophysiological correlates of organizational change and threat of unemployment among police inspectors. *Integrative Physiol and Behav Science* 1999; 34: 30–42.

Gustavsen B, Hunnius G. *New Patterns of Work Reform: The case of Norway*. Oslo University Press, Oslo, 1981.

Hasselhorn H-M, Theorell T, Vingård E, the MUSIC Study Group. *Acute Musculoskeletal Disease and Psychophysiological Parameters – A Swedish Case Control Study*. Stress Research Report. Stress Research Reports Nr 289, ISSN 0280-2783 National Institute for Psycho-Social Factors and Health, PO Box 230, 17177 Stockholm, Sweden, 1999.

Holmström EB, Lindell J, Moritz U. Low back and neck/shoulder pain in construction workers: Occupational workload and psychosocial risk factors. *Spine* 1992; 17: 672–677.

Johansson H, Sjölander P. Neurophysiology of joints. In *Mechanics of Human Joints,* V Wright and EL Radin, eds. Dekker, New York, 1993.

Johansson JÅ, Rubenowitz S. *Arbete och besvär i nacke, skuldra och rygg.* Mimeograph. Inst. of Psychology, University of Göteborg, 1992.

Johnson JV, Hall EM. Job strain, workplace social support and cardiovascular disease: A cross-sectional study of a random sample of the Swedish working population. *Am J Public Health* 1988; 78: 1336–1342.

Josephson M. Work factors and musculoskeletal disorders – An epidemiological approach focusing on female nursing personnel, Thesis. Karolinska Institutet, Stockholm, Sweden, 1998.

Karasek RA. Job demands, job decision latitude, and mental strain: Implications for job redesign. *Admin Sci Q* 1979; 24: 285–307.

Karasek RA, Theorell T. *Healthy Work*. New York: Basic Books, 1990.

Kamwendo K, Linton S, Moritz U. Neck and shoulder disorder in medical secretaries. *Scand J Rehab Med* 1991; 23: 127–133.

Kopelman RE. Job redesign and productivity: A review of the evidence. *Nat Productivity Rev* 1985 (Summer); 237–255.

Leino PI, Hanninen V. Psychosocial factors at work in relation to back and limb disorders. *Scand J Work Environ Health* 1995; 21(2): 134–142.

Linton SJ. Risk factors for neck and back pain in a working population in Sweden. *Work and Stress* 1990; 4: 41–49.

Linton SJ, Bradley LA. An 18-month follow-up of a secondary prevention program for back pain: help and hindrance factors related to outcome maintenance. *Clin J Pain* 1992; 8(3): 227–236.

Lundberg U, Kadefors R, Melin B, Palmerud G, Hassmén P, Engström M, Elsberg Dohn I. Psychophysiological stress and EMG activity of the trapezius muscle. *Int J Behavioral Med* 1994; 1(4), 354–370.

Maier SF, Dragan RC, Oran JW. Controllability, coping behaviour and stress-induced analgesia in the rat. *Pain* 1982; 12: 47–56.

Orth-Gomér K, Eriksson I, Moser V, Theorell T, Fredlund P. Lipid lowering through stress management. *Int J Behav Med* 1994; 1: 204–214.

Pietri F, Leclerc A, Boitel L, Chastang J-F, Morcet J-F, Blondet M. Low-back pain in commercial travelers. *Scand J Work Environ Health* 1992; 18: 52–58.

Reesor K, Craig KP. Medically incongruent chronic back pain: Physical limitation, suffering and ineffective coping. *Pain* 1987; 32: 35–45.

Rundcrantz B-L, Johnsson B, Moritz U, Roxendal G. Occupational cervico-brachial disorders among dentists. *Scand J Soc Med* 1991; 19: 174–180.

Schnall PL, Landsbergis PA. Job strain and cardiovascular disease. *Ann Rev Public Health* 1994; 15: 381–411.

Silverstein B. Design and evaluation of interventions to reduce work-related musculoskeletal disorders. Lecture at PREMUS, Stockholm 1992. *Arbete och Hälsa* 1992; 17: 1–7.

Theorell T, Flodérus B, Lind E. The relationship of of disturbing life-changes and emotions to the early development of myocardial infarction and other serious illnesses. *Int J Epidemiol* 1975; 4: 281–296.

Theorell T, Nordemar R, Michélsen H, Stockholm MUSIC Study Group. Pain thresholds during standardized psychological stress in relation to perceived psychosocial work situation. *J Psychosom Res* 1993; 37: 299–305.

Theorell T, Harms-Ringdahl K, Ahlberg-Hultén G, Westin B. Psychosocial job factors and symptoms from the locomotor system – a multicausal analysis. *Scand J Rehab Med* 1991; 23: 165–173.

Theorell T, Karasek RA, Eneroth P. Job strain variations in relation to plasma testosterone fluctuations in working men – a longitudinal study. *J Intern Med* 1990; 227: 31–36.

Theorell T, Alfredsson L, Westerholm P, Falck, B. Coping with unfair treatment at work - how does the coping pattern relate to risk of developing hypertension in middle-aged men and women. *Psychother Psychosom* 2000; 69: 86–94.

Theorell T, Perski A, Åkerstedt T, Sigala F, Ahlberg-Hultén G, Svensson J, Eneroth P. Changes in job strain in relation to changes in physiological state. *Scand J Work Environ Health* 1988; 14: 189–196.

Theorell T. Possible mechanisms behind the relationship between the demand-control-support model and disorders of the locomotor system. *Beyond Biomechanics: Psychosocial Aspects of Musculoskeletal Disorders in Office Work,* in SD Moon, SL Sauter, eds. Taylor and Francis, London, 1996.

Theorell T, Orth-Gomér K, Moser V, Undén A-L, Eriksson I. Endocrine markers during a job intervention. *Work and Stress* 1995; 9: 67–76.

Theorell T, Tsutsumi A, Hallquist J, Reuterwall C, Hogstedt, C., Fredlund P, Emlund N, Johnson J, Stockholm Heart Epidemiology Program (SHEEP). Decision latitude, job strain, and myocardial infarction: A study of working men in Stockholm. *Am J Public Health* 1998; 88: 382–88.

Theorell T, Hasselhorn H-M, Vingård E, Andersson B, the MUSIC-Norrtälje Study Group. Interleukin 6 and cortisol in acute musculoskeletal disorders – results from a case-referent study in Sweden. *Stress Med* 2000; 16: 27–35.

Tola S, Riihimäki H, Videman T, Viikari-Juntura E, Hänninen K. Neck and shoulder symptoms among men in machine operating, dynamic physical work and sedentary work. *Scand J Work Environ Health* 1988; 14: 299–305.

Toomingas A, Theorell T, the Stockholm MUSIC Study Group. On the relationship between psychosocial working conditions and symptoms/signs of locomotor disorders in the neck/shoulder region. Lecture at the PREMUS, Stockholm May 1992.

Wahlstedt K, Nygård CH, Kemmlert K, Torgén M, Gerner Björkstén M. Påverkan av en organisationsförändring på arbetsmiljöfaktorer och upplevd hälsa inom brevbäring, Stockholm 1996. *Arbete och hälsa* 1996; 15: 1–29.

Wallin L, Wright I. Psychosocial aspects of the work environment: A group approach. *J Occ Med* 1986; 28: 384–393.

Vingård E, Alfredsson L, Hagberg M, Kilbom Å, Theorell T, Waldenström M, Wigeaus-Hjelm E, Wktorin C, Hogstedt C, the MUSIC-Norrtälje Study Group. *Spine* 2000; 25: 493–500.

Wigeaus-Hjelm E. The influence on neck and shoulder disorders from work-related physical and psychosocial exposure among men and women in a Swedish general population. *Abstract 14th Int. Conf. on Epidemiology in Occ. Health*, Herzliya, Israel, 10–14 October 1999.

Chapter 5

Analysis and Design of Jobs for Control of Work Related Musculoskeletal Disorders (WMSDs)

Thomas J. Armstrong

... and now remains
That we find the cause of this effect –
Or rather say, the cause of this defect,
For this effect defective comes by cause.

Polonius in *Hamlet*, Act II Scene ii

5.1 INTRODUCTION

It has been shown in previous chapters that work related musculoskeletal disorders, (WMSDs) are a leading cause of disability and workers compensation. This chapter is concerned with analysis and design of jobs for both primary and secondary prevention of disability associated with WMSDs. Particular emphasis is given to the analysis as it provides an understanding of the cause of the exposure and insight into those aspects of the job that should be redesigned to eliminate or at least reduce that exposure. This chapter focuses on physical analysis and design of physical risk factors. Psychosocial factors are discussed in Chapter 4.

5.2 ANALYSIS OF JOBS

Job analysis is a systematic process for obtaining information about a job. In this case the desired information is concerned with what the worker does and that exposes him or her to WMSD risk factors. This section describes the information needed to characterize what the worker does, sources of information for determining what the worker does, the relationship between what the worker does and his or her exposure to physical risk factors of WMSDs and finally recognition and quantification of risk factors.

5.2.1 What the Worker Does

Collection of information about what the worker does is referred to as job documentation and it draws heavily on field work measurement and can be traced back to antiquity. For this discussion, the reader is referred to the classic work of Gilbreth (1915) or the more contemporary works of Barnes (1980) and finally the recent work of Niebel and Freivalds (1999). The basic information included in documentation of the job is summarized in Table 5.1 and is intended as a guide that can be used for the first step of the job analysis.

Table 5.1. Information that should be recorded as part of the job documentation to determine what the worker does.

Job Documentation	
1. Formal job title	
2. Descriptive job title	
3. Work objectives (List sub-objectives in part B)	
4. Production/work standard	
Repeat the following for each sub-objective (Task) fill out 5 & 6 only if 2 or more tasks	
5. Title	
6. Production/work standard	
7. Objective	
8. Approximate time task is performed (minutes, hours, %)	
9. Work station location with key physical dimensions and environmental attributes	1) 2) 3) 4)
10. Tools and equipment (including personal protective equipment) with key dimensions, weights and force requirements	1) 2) 3) 4)
11. Materials with key dimensions & weights	1) 2) 3) 4)
12. Element attributes	

n	Element	Work object	Time	Reach x, y, z	Gross posture	F, lift/ push	Hand posture	F, grip/ pinch	Feel	See	Hear	Decide
1												
2												
3												
4												
5												
6												
7												

The job title

Job titles used by employers tend to be very general, e.g., line worker, assembler or clerk. In other cases the name may refer to a specific machine or part, e.g. 3141 Operator. In either case, these names do not provide much insight into the content of the job. Recording the job title will make it possible for the analyst to identify the specific job for the employer. The analyst should also record a descriptive job title, e.g. small parts assembler, word processing operator for health care providers, for workers and others not specifically familiar with the workplace. Descriptive title provides insight into what the worker does. For example, the job title 'clerk' implies that the worker might use a computer for data entry, but the descriptive title 'medical transcription' implies that data entry and word processing are large parts of the job.

The purpose or objective

Basically the objective is a statement of why the job exists. It is something like the descriptive job title, but it is more specific. As a minimum the objective includes a verb and a noun, but in some cases a few additional words can greatly increase the quantity of information. An extensive list of verbs can be found in Birt et al. (1996). Several examples are shown in Table 5.2. For example 'install batteries in car' or 'install chips in PC board' provide more information than 'install parts.' Objectives should not be too wordy, e.g., accurately assemble parts. It can generally be assumed that quality is important. Specific production and quality information, along with materials and tools, are documented in subsequent steps of the analysis.

Table 5.2 Simple work objectives typically consist of a verb and a noun and provide limited insight into what the worker does. Descriptive objectives typically consist of a phrase that provides insight into the equipment and actions used.

Simple objective	*Descriptive objective*
Operate press	Setup and monitor automatic press, manually load and operate small press
Install parts	Install batteries in car, install small parts into transmission assembly, install circuit board components
Identify defects	Identify bugs in software, identify blemishes in painted parts
Enter data	Transcribe dictation, enter medical billing data into computer

In some cases jobs include multiple objectives. All sub-objectives should be listed; some examples include:

Homemaker:

- Prepare meals
- Keep house clean
- Maintain stock of clean laundry
- Maintain food and household supplies

Clerical worker:

- Type/edit letters and reports from handwritten drafts
- Make copies as needed
- Answer and refer phone inquiries
- Send, get and sort post

Press operator:

- Get and stock materials
- Load/operate press

Each sub-objective is like an additional job, although they may be interspersed at regular or irregular intervals throughout the work shift. The activities that correspond to a sub-objective are referred to as a task. Jobs consist of one or more tasks. Each task should be documented separately. The effort and time required to document and analyse a job increase with increasing number of tasks.

Production or work standard

The production standard is a statement of how much work is to be performed in a given amount of time. In some cases the standard is a formal statement of parts per hour or parts per shift, which may be expressed in terms of average assembly line speed. Production standards generally take into consideration time for process delays and personal needs. Standards may include an incentive system that compensates workers for producing more than the base standard. In some cases workers may be allowed to rest when they complete their basic work quotas. These details should be documented for consideration in later steps of the analysis.

In addition to the production standard, it is necessary to document the work procedures or 'methods' by which that standard is achieved. The method is a sequence of steps or work 'elements' along with corresponding equipment and materials. Gilbreth (1915) proposed a list of 'elements' or 'Therbligs' for describing tasks that is still widely used in various forms. As an aside, Gilbreth's paper describes the use of work methods analysis for identifying barriers to employment of disabled soldiers. All of the Therbligs are verbs that describe the body doing something, e.g. 'Grasp' where the hand applies force to embrace a work object, 'Move' where the hand applies force to transfer something from one location to another, 'Reach' where the hand moves to obtain or touch an object, etc. Figure 5.1 shows a job in which a worker loads and operates a small press. The method used to load and activate the press is as follows:

Load and Activate Press

(1) Reach (with tongs) for part
(2) Grasp part
(3) Move part to hold position
(4) Move part to press
(5) Position part in press
(6) Reach for controls (both hands)
(7) Apply force to activate press (press automatically ejects part)

The level of detail can be increased or decreased as necessary; however, it is generally desirable to keep it simple and detail only as necessary. Load and activate press job above, can be simplified as:

Load and Activate Press

(1) Get part and load press
(2) Operate press

Lists of standard verbs can be found in Birt et al. (1996).

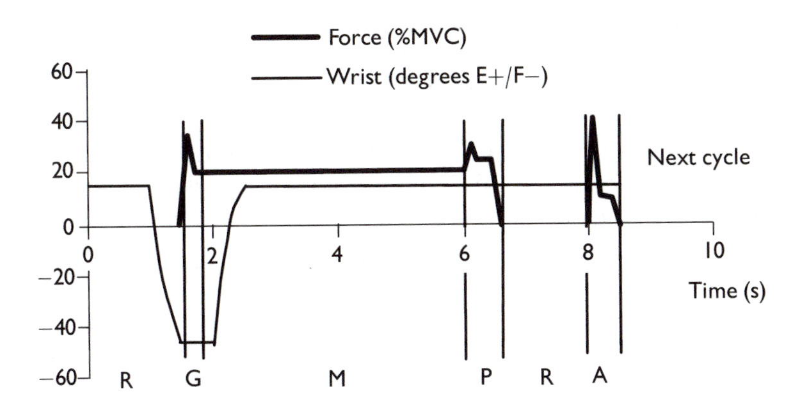

Figure 5.1 Finger force (%Maximum Voluntary Contraction) and wrist postures (degrees) versus time for a worker operating a small press. (R = Reach, G = Grasp, M = Move, P = Position, A = Activate).

The production standard and method are unique to a particular work setting, but in some cases it can be generalized to other settings. It is important that the analyst observe the job to make sure that the official work method corresponds with what the workers actually do.

The production standard may not be formally stated; however, there will still be an expectation of how much work is to be performed. For an insurance claim processor it may be an average number of claims per day, but because of variations in the nature of claims, the actual number may vary from day to day and worker to worker. In other cases the expectation may vary from project to project, e.g. software developer, equipment designer, construction contractor. Many jobs have specific quality specifications, e.g., food preparation, parts finishing, data entry, etc. In other cases quality specifications may be implicit. It can be assumed that if a worker puts a part into a machine it should be done correctly and need not be stated. These requirements should be documented.

Task attributes

A task is defined as the activities associated with a sub-objective of a job. The attributes of each task that affect worker demands or exposure to musculoskeletal disorder risk factors should be recorded. There is more information than can be recorded in a reasonable amount of time. Also there is more information than is really useful. With experience, the analyst will gain insight into which information should be recorded.

The analyst should consider why they are analysing the job in the first place. Possible reasons might include the following:

(1) A worker has been injured and information is needed to determine if work modifications are necessary for that worker to return to their job and prevent future injuries. If the case involves a hand/wrist injury, the analysis will focus on the hand activities. If the injury involves the shoulder, the analysis will focus on shoulder activities.
(2) An employer is introducing a new product line and the analysis is intended to identify possible problems with existing designs so that they are not replicated in the new jobs. In this case, the analysis will need to be quite detailed and consider all possible problems and workers.

The task analysis requires information about the workstation, equipment, tools and materials.

Workstation

Workstations may vary considerably from one job to the next, e.g., a worker sitting at a desk, a worker standing at a work bench, a worker walking alongside a car as it moves down a production line, a hairdresser standing next to a seated client, a lineman strapped to a utility pole, etc. The workstation should include a name and key dimensions. Key dimensions are the work locations where the worker must reach with one or both hands. The dimensions should be measured with respect to the barriers that determine where the worker will be located. The vertical dimension is typically measured with respect to the floor, but in some cases will be the vertical limits of a seat. In some cases, such as where the hairdresser can raise or lower their client, or the lineman that can move up or down the utility pole, the vertical dimension may not be constrained and only the limits would be listed. The horizontal dimensions are typically measured with respect to barriers or obstructions that limit how close the worker can be to the machine, e.g. front of the machine or front edge of the workbench. Similarly, the lateral dimension would be measured with respect to constraints that affect the worker s movement side to side. Often the major lateral restraint is the location of a seat.

The workstation dimensions may be reported with a drawing, in a table or in text. Figure 5.2 shows a workstation layout for a worker at a desk. The drawing need not capture all the details of the workstation unless they are directly relevant to what the worker does.

The workstation documentation should indicate the presence of environmental factors that might affect the worker s performance, e.g. lighting, noise or vibration. Additional analyses of environmental factors require specialized equipment and go beyond the scope of this discussion.

Equipment

Equipment includes machines, hand tools, mechanical assists, computers, computer accessories, software and personal protective equipment. Equipment attributes include key dimensions that affect how a worker positions and moves their body to complete the task. If the equipment were held, the shape, weight and reaction forces would be important.

Materials

Materials include things that become part of the product or a by-product of the process. Examples include: raw materials, parts, fasteners, documents, references, etc. Key material attributes include the size and weight as they affect how the worker would hold and use the material.

Work element attributes

The following information should be included for each work element: The work object (tools and materials), the duration and frequency of the element, the *x*, *y* and *z* location of the work objects, gross body posture (lift, pull or

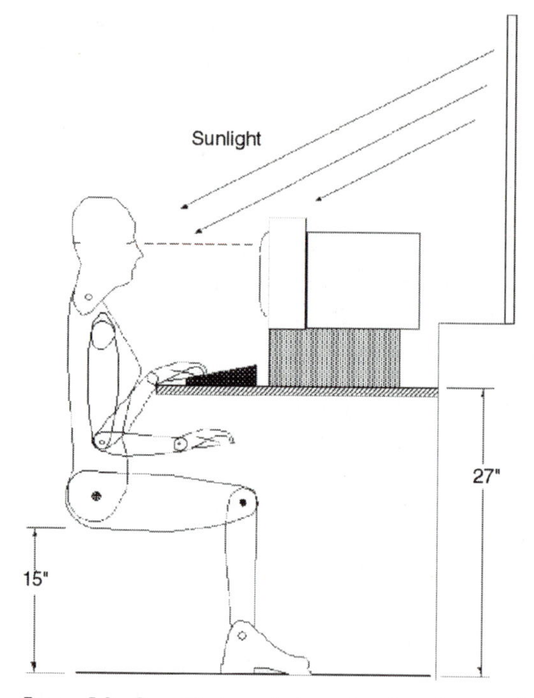

Figure 5.2a Simplified workplace drawing showing key dimensions and environmental factors. Inclusion of a manikin based on fifth percentile female stature, it can be seen that the worker will have to reach up to the keyboard.

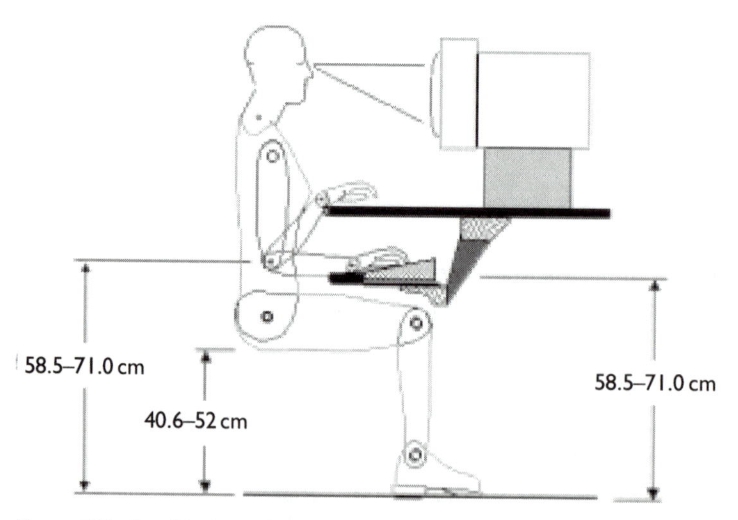

Figure 5.2b Simplified workplace drawing showing how an adjustable keyboard tray reduces worker exposure to posture and contact stresses for a small worker.

push), hand force, hand posture (pinch or grip), the sensory requirements (feel, see or hear) and decisions (see Table 5.1). These attributes should be self-explanatory. It may not be necessary to collect all of this information, depending on the reason for doing the analysis in the first place.

5.2.2 Sources of Information for Job/Task Documentation

Job information generally is obtained from four sources: existing job documentation, inspections of the worksites, interviews of workers and supervisors, and observations. This section describes use of these sources to complete the job documentation.

Existing Documentation varies considerably from one workplace to another. Manufacturing sites often have explicit production standards expressed in terms of work quantities per unit of time. These standards may be based on time study, predetermined time study data or experience. Time study is a method of determining the time required for a qualified and trained worker to complete the job under normal circumstances (Barnes 1980; Niebel and Freivalds 1999). It is worth reviewing the steps of a time study because it is an excellent source of job data and the identification of ergonomic stresses draws heavily on time study methods. Time study can be divided into the following steps:

(1) documenting the workstation and method
(2) selecting a normal operator, conducting the time study
(3) analysing the data to determine the average time
(4) adjusting the average time according to normal expectation
(5) adjusting the normal time for process delays and personal needs (e.g. fatigue)

Predetermined time systems, PDS, provide a mechanism for determining time standards before jobs are set up (Barnes 1980; Niebel and Freivalds 1999). This makes it possible to balance work allocations and estimate the labor cost of new products before going into production. Application of predetermined time systems entails the following steps:

(1) describing the workstation, materials, tools and methods
(2) applying empirical time equations for specific design parameters of that job, e.g. presentations of materials and work objects, reach and move distances, part sizes, fit tolerances of assemblies, etc. to estimate the time required to perform each work element for each hand. In the past these data were tabulated for different conditions to facilitate their application. Increasingly, computers are used to perform these calculations.
(3) calculating the total time for a complete work cycle
(4) adjusting the normal time for process delays and personal needs (e.g. fatigue)

Predetermined time systems provide detailed information about the job. Some systems, notably ErgoMOST®, include specific consideration for ergonomic parameters (Zandin et al. 1996). Ergonomic issues are best dealt with at the time jobs and equipment are designed.

Production standards are often based on experience. In some cases employers have a formal system for collecting production data and relevant job/process parameters. Statistical models then are used to generalize these 'standard data' to new work situations. In other cases production may be based on 'common knowledge' or 'rules of thumb'. Just-in-time and lean manufacturing often require employers to reconfigure production processes on short notices and to achieve high levels of efficiency. This has resulted in increased use of formalized predetermined time systems and standard data.

Other documentation includes existing workplace and equipment specifications. Many of the newer manufacturing sites are likely to have been laid out using computerized drafting systems. These drawings provide important information about the spatial relationship between work objects and workers. Workplace drawings should always be compared with observations, because workplace modifications may not be documented. 'Off-the-shelf' equipment, e.g. seating, hand tools, typically comes with documentation describing its performance specifications and its use. These documents can be used to determine weights of parts, ranges of adjustability, torque settings, etc. Some equipment is designed specifically for a given operation and documentation may not be available.

Observations are one of the principle methods used for documenting what the worker does and identifying ergonomic stresses. It is important that observations be conducted in such a way that they (1) do not interfere with the job and (2) provide a representative sampling of what occurs over the course of the hour, shift, season and year. The purpose of the observations should be explained to the workers. That also is a good opportunity to ask questions about what they do and to obtain their consent for photographs or videotaping. Observations should be taken in a way that does not interfere with the worker or create a safety problem. After initial introductions, the analyst may rotate between several workstations to see if and how the jobs change over time.

Videotapes provide a permanent record of work activities and facilitate job comparisons. It is important that several representative cycles of the job be recorded. Recordings also should capture normal work variations. Many camcorders can be programmed to take a one- or two-second video clip every 30 to 60 seconds. Real time video clips are superior to single frame recordings because they make it possible to determine if the hands are moving or resting.

The fraction of time, p_i, that a given event, i, occurs is calculated as the number of times that event is observed divided by the number of observations. Because of variations in the process and the observations, the value of p_i can be expected to vary from one set of observations to another. If the observations are random, a confidence interval for p_i can be calculated. Generally there will be enough randomness between the process and the observations that a normal approximation can be used to describe the statistical distribution of thirty or more observations. The 95% confidence limit for the fraction of time, p_i, for a given event, i, is calculated as:

$$p_i - 1.96[p_i(1 - p_i)/n]^{1/2} \leq p_I \leq p_i + 1.96[p_i(1 - p_i)/n]^{1/2}$$

Ninety-five percent confidence limits for events that occur 10%, 30% and 50% of the time based on 30, 60, 120, 240 and 480 observations are shown in Table 5.3. It can be seen that the confidence limit decreases with increasing sample size and decreasing p_i (Niebel and Freivalds 1999).

Table 5.3 Confidence limits for events that occur 10%, 30% and 50% of the time based on 30 to 480 random observations.

N	P = 0.01	P = 0.3	0 = 0.5
30	0–0.21	0.14–0.46	0.32–0.68
60	0.02–0.18	0.18–0.42	0.37–0.63
120	0.05–0.15	0.22–0.38	0.41–0.59
240	0.06–0.14	0.24–0.36	0.44–0.56
480	0.07–0.13	0.26–0.34	0.46–0.54

A word of caution is in order regarding the intervention described in the packing example. There is often a great temptation to increase the productivity and pay for the intervention. If that were to happen, the repetition might not be changed. The only gain in that case would be a reduction in force. The employer may need to be educated about risk factors of WMSDs and that the payback will be in reduced injury and illness cases.

The events of interest may pertain to specific stresses, e.g. working with an elevated shoulder posture or use of a particular device such as a hand tool or keyboard. Figure 5.3 shows a bar chart of the percent time devoted to different tasks for an insurance claims processor. In this case there was specific interest in use of the keyboard. It can be seen that the keyboard is used only about 9.8% of the time. Since this estimate was based on 297 observations, the 95% confidence limit is from 6.4% to 13.2%.

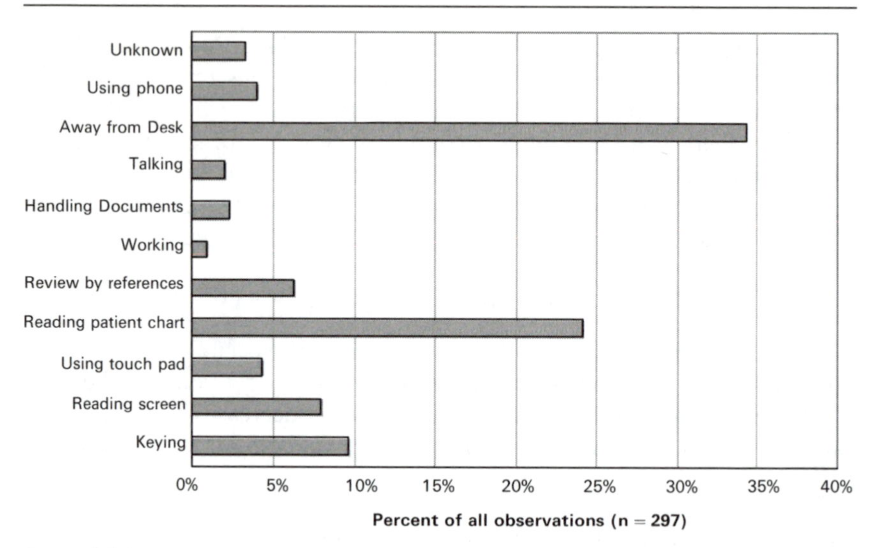

Figure 5.3. Distribution activities for medical billing job based on 297 one-second video recordings of the workstation (1/min).

Worker and Supervisor Interviews

Workers and supervisors are important sources of information about the job. Interviewer techniques can greatly influence the quality of information obtained (McCormick 1984; Sinclair 1995). Some key points for the interview include:

- The interview should have a clearly stated purpose.
- The interviewer should not allow the interview to wander from the stated purpose.
- The interviewer should avoid using responses that express approval or disapproval, e.g. 'that's good' or 'too bad'. Neutral terms should be used, e.g. 'I see'. In some cases it is helpful to summarize the worker s statements to clarify his or her meaning.
- The interviewer should avoid asking leading questions, e.g. 'What is it about this job that makes your shoulder hurt?' If the purpose of the interview is to investigate the cause of a worker's pain, it is better to say 'Are you aware of any activities in your life that are particularly painful to your shoulder? Please list and describe them.'

Interviews are often facilitated by using a checklist to make sure that all of the basic points of the interview are covered.

Homan et al. (1999) has reported differences as much as two to one between workers perception of how much of the time they use keyboards versus video recordings of keyboard use. This difference may reflect differences between the workers' perception and the observer's about what constitutes keyboard use. If a worker stops keying to read a message on the screen, they may perceive this as continuing to use the keyboard. To the analyst, their hands are now resting and they are not exposed to the stresses of keying. Questions about equipment use and job stresses need to be very carefully crafted.

The source of information will determine the reliability of the analysis. It should not be assumed that data based on interviews are inherently inferior to work site inspections or physical measurements. Each source has distinct advantages and disadvantages. Most analyses will draw on information from multiple sources.

5.2.3 Identifying Physical Stresses

The term 'Work-Related Musculoskeletal Disorder,' or WMSD, refers to a group of disorders that involves repeated or sustained loads on the musculoskeletal system. Affected tissues, mechanical factors and physiological responses differ from one disorder to another. WMSDs may be influenced by psychosocial factors. Several models have been put forward to explain the relationship between these factors at conceptual and operational levels (Armstrong et al. 1993; Bongers et al. 1993; Kilbom 1994ab; McAtamney and Corlett 1993; Moore and Garg 1995). For the purposes of this discussion, an overview will suffice.

It has already been shown that a job can be characterized as a sequence of steps or elements. These elements, along with the attributes of the workplace, equipment, materials and the individual worker determine the forces on the body and the postures that must be assumed to complete the task. Finger force and wrist flexion/extension angles are shown as functions of time in Figure 5.1 for a job in which a worker gets a part, puts it into a machine and activates the machine control. Unless constrained by the design of work equipment, workers generally will maintain a slightly extended wrist position. In this case the worker must 'Reach' into bin and 'Grasp' the part (see Figure 5.4). The location and orientation of the bin with respect to the seated worker results in 45° of wrist flexion. The worker is able to maintain 15° wrist extension for the other work elements since posture is not constrained by the workstation layout.

Finger force is related to size, shape, weight and friction of the work object. In this case the part is most easily grasped with the fingertips in a pinch posture. Also, a pinch grip is used to position the part in the machine. The force required to hold the part is related to the weight of the object and its surface friction with respect to the worker s gloves or skin. Figure 5.5a shows a free body diagram of the fingers holding a work object against the force of gravity. The required pinch force, F_p, is equal to the weight of the part, W, divided by two times the coefficient of friction, μ.

Figure 5.4 Worker must flex left wrist to get part from bin.

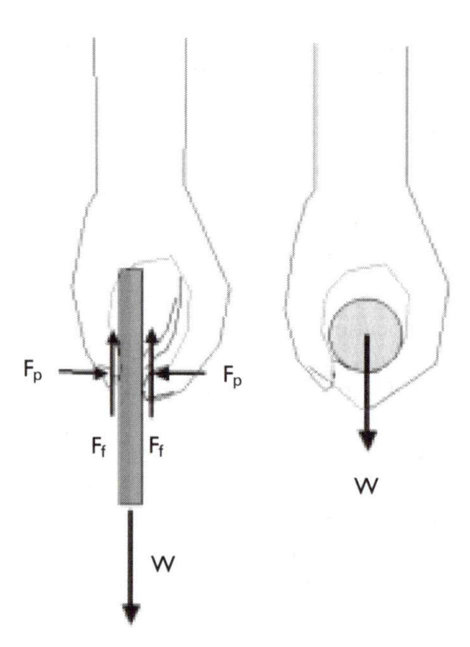

Figure 5.5 Free-body diagram of the fingers holding a work object against the force of gravity using (a) pinch grip and (b) hook grip. For pinch grip, the force required to hold the object is affected by surface friction.

$$F_p \geq W/2\mu$$

Parts can be weighed and coefficients of friction for skin have been reported by Buchholz et al. (1988) and Bobjer et al. (1993). In this example, the part has a weight of 6N and the coefficient of friction is equal to 0.5.

$$F_p \geq 6/(2 \times 0.5) = 6\text{N}$$

Forces are usually converted to percent of maximum voluntary contraction or %MVC for an individual. A 6N pinch by a worker with 30N pinch strength is equal to 20%MVC. This is the minimum force to keep the object from slipping out of the fingers. It has been shown that people generally exert more than the minimum required force especially at the beginning of the exertions (Frederick and Armstrong 1995; Westling and Johansson 1984; Armstrong et al. 1994). This is particularly true for work elements such as 'activate machine' in which the worker uses ballistic movements to strike the machine controls.

Exertion of the hand involves mechanical equilibrium between the external load forces and the internal inertial, viscous and elastic forces of the forearm and hand. As a result certain postures will result in greater stresses on certain tissues than others. These equilibrium forces result in elastic and viscous deformation of tissues and in increased carpal tunnel pressure (Goldstein et al. 1987; Lundborg et al. 1982; Werner et al. 1997; Rempel et al. 1997, 1998). In addition, increased metabolic demand is required to maintain muscle force. Prolonged or repeated tissue deformation, increased compartmental pressure and metabolic demand can lead to tendon, nerve and muscle damage.

If the tendons are perfectly aligned with the muscle and its point of attachment, then the dominant tendon forces will be parallel to the long axis of the tendon. Anatomical structures, e.g., bones or ligaments, that protrude into the tendon path, result in forces acting perpendicular between the tendon and adjacent structures (Armstrong and Chaffin 1979). A schematic diagram of the finger flexor muscles and tendons is illustrated in Figure 5.6. If there is movement, e.g., flexion or extension of the wrist, which causes the tendons to slide back and forth, then there also will be shear or friction forces acting parallel between the tendon and adjacent structures. With flexion of the wrist, not only are the tendons pressed against the adjacent palmar carpal ligaments, but they also are pressed directly against the median nerve. The Phalen s test for carpal tunnel syndrome utilizes this concept to evaluate the sensitivity of the median nerve to mechanical compression (Phalen 1966). The mechanics of the tendons stretched around an adjacent surface are analogous to a belt stretched around a pulley (Armstrong and Chaffin 1979). It can be shown that the contact stress, σ, is related to tendon tension, F_t, joint angle, θ, tendon friction, μ, radius of curvature, r, and width of the tendon, w:

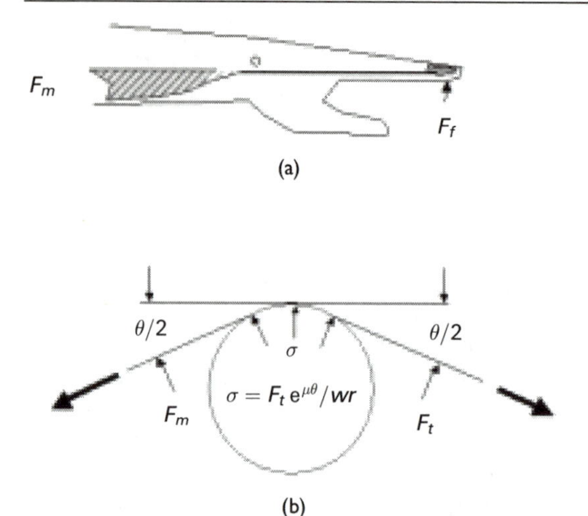

Figure 5.6 (a) Exertion of finger force, F_f, requires muscle force, F_m, to maintain static equilibrium. (b) Flexion or extension of the wrist will cause the tendons to press against the internal structures of the wrist, producing a contact stress, σ, perpendicular to the structure and the tendon.

$$\sigma = F_t e^{\theta\mu}/rw$$

The friction forces are quite low and insignificant in healthy tendons. When the tendon becomes irritated, the friction increases and may interfere with joint motion.

If friction is very low, the muscle force, F_m, and tendon force, F_t, will be equal and the above equation for contact stress can be simplified to:

$$\sigma = F_t/rw$$

A pinch force of 10 N, which may require as much as 30 N of tendon force will result in 0.4 MPa (approximately 3,000 mm Hg) for a wrist with a radius of curvature $r = 15$ mm, and a tendon width $w = 5$ mm. This pressure is great enough to interfere with biological processes and account for histological changes commonly reported in the radial and ulnar bursa, as well as the median nerve inside the carpal tunnel.

In addition to producing contact between the tendons and internal structures, flexion/extension of the wrist results in increased intercarpal pressure. Empirical measurements show that these pressures increase with wrist deviation from neutral and with increasing force exertions (Werner et al. 1997; Rempel et al. 1997, 1998). Intercarpal pressures vary significantly from person to person, but may exceed 70 mm Hg. These pressures have been shown to interfere with blood flow and nerve function.

Exertions also impose metabolic demands on muscles. If the demand exceeds the steady state capacity of the muscle, anaerobic metabolism will occur. Aerobic metabolism cannot sustain the exertions and eventually metabolic substrates will be depleted and metabolites will be accumulated to the point where the worker experiences discomfort and impaired motor control. Eventually the worker may be unable to continue the exertion. In extreme cases there even may be muscle damage (Armstrong et al. 1993; Bystrom and Fransson-Hall 1994).

The body can adapt to some of the effects of mechanical stress and high metabolic demands over time. If the stress is too great or if there is not sufficient recovery time between repeated exertions, a muscle, tendon or nerve may become impaired. In some cases these impairments are injurious and work capacity is decreased. In other cases they are adaptable and work capacity is increased. In some cases the adaptations of one tissue may adversely affect another. For example, hypertrophy of the radial and ulnar bursa surrounding the flexor tendons in the carpal tunnel reduces the available space and contributes to secondary compression of the median nerve (Phalen 1966). Connective tissue appears to develop in response to the mechanical stress within the median nerve itself (Armstrong et al. 1984).

Symptoms may develop before there is clinical evidence of a lesion or conspicuous impairment. Also, the clinical procedures for identifying impairments may be more traumatic than the impairment itself. As a result, most health care providers will intervene based on workers reports of symptoms. Similarly it can be argued that jobs should be designed to minimize risk of symptoms that will result in visiting a health care provider.

It has been shown that exertions and postures of the hand are related to work procedure and design specifications. Furthermore, exertions and postures of the hand are linked to loads and impairments of muscles, tendons and nerves in the wrist and forearm. Similar relationships can be shown for other parts of the body, but are beyond the scope of this discussion. These relationships provide a basis for (1) identifying work design parameters and work activities associated with musculoskeletal disorders; and (2) for designing and interpreting epidemiological studies.

5.2.4 Recognizing and Quantifying WMSD Risk Factors

Recognized factors of musculoskeletal disorders and symptoms include:

- Repeated and sustained static exertions
- Force
- Contact stresses
- Postures
- Low temperature
- Vibration

This section describes the qualities of each risk factor that can be identified from the job documentation described above. After the factors are identified, methods can be selected for obtaining the desired level of quantification. Methods for quantifying exposure include: worker ratings, observer ratings, biomechanical analysis, and instrumentation.

Repeated and sustained static exertions refer to the temporal pattern of forces and postures. These patterns result in mechanical and physiological loads on the musculoskeletal system. They are often characterized by exertion time, recovery time, cycle time and exertion frequency. These metrics can be estimated from elemental analysis such as that shown in Figure 5.1. All of these metrics have been used in laboratory and epidemiological studies; however, we find cycle time by itself can be misleading. Job A shown in Figure 5.7 involves two 3 s exertions and a 10 s cycle time; Job B involves six 3 s exertions and a 30 s cycle time. Cycle time alone gives the impression that the second job is less stressful that the first; however, the exertion frequency, exertion time and recovery time are the same for both jobs. Job C also has a 30 s cycle time, but requires only two 3 s exertions because the operator must wait for the machine to cycle. The worker performing job B will be resting 40% of the time while the worker performing job C will be resting 80% of the time. If everything else is the same, the worker performing Job B will be exposed to much greater stress than the one performing Job C. In addition to cycle time, knowledge of the work content is necessary so that exertion frequency and recovery time can be computed.

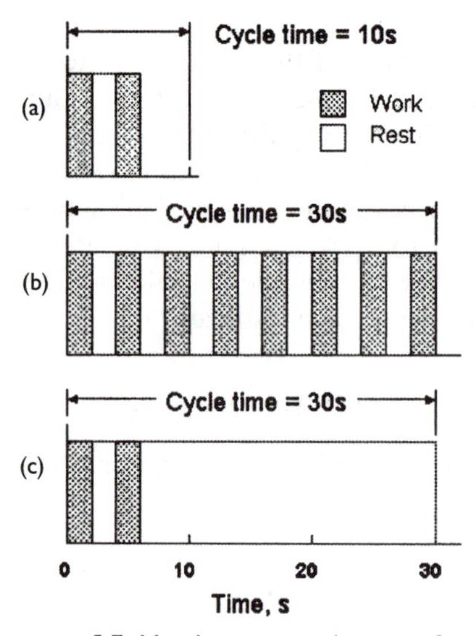

Figure 5.7 Hand repetition/activity for three jobs. Job A requires 2–3 s exertions and a 10 s cycle time; Job B requires 6–3 s exertions and a 30 s cycle time; and Job C requires 2–3 s exertions and a 30 s cycle time.

Repetition can also be rated from job observations and/or video recordings of representative jobs (Latko 1997). A zero to ten scale for rating repetition is shown in Figure 5.8. We prefer to have two or more people rate jobs independently and then discuss their ratings to achieve consensus. These ratings have been shown to have a close relationship with exertion frequency and recovery time. Symptoms of certain upper limb musculoskeletal disorders, including carpal tunnel syndrome, tendon disorders and non-specific pain were all found to increase linearly with increasing repetition ratings from 2.4 to 8.0 for industrial workers (Latko et al. 1999). The ACGIH recently adapted this metric along with normalized force levels as a basis for a threshold limit value, TLV, for mono-task handwork (ACGIH 2000).

Instrumental methods for measuring repetition include the use of electromyography (EMG) to record extrinsic forearm muscle activity and goniometers to measure wrist velocity and acceleration. These methods examine the duration and frequency of forces and movements. Specific examples of these include amplitude probability distributions proposed by Jonsson (1988) and exposure variation analysis (EVA) proposed by Mathiassen and Winkel (1991) for summarizing muscle activity. Marras et al. (1993) reported that wrist velocity and acceleration measured with an electromechanical goniometer were associated with OSHA reported repeated trauma disorders.

Commercial goniometers are now available that can be attached to a joint to provide an electrical signal proportional to joint position. The signal is generally recorded on the computer and summarized as a frequency histogram and corresponding descriptive statistics. Posture for repetitive work has periodic properties that can be characterized using a series of sine. Velocity and acceleration are calculated as the first and second time derivatives of posture. Posture, velocity and acceleration functions can be written as:

$$\text{Posture} = \sum \theta_{pi} \sin(2t\pi/T_i + \Phi_i)$$

$$\text{Velocity} = \sum \theta_{pi}(2\pi/T) \cos(2t\pi/T_i + \Phi_i)$$

$$\text{Acceleration} = -\sum \theta_{pi}(2\pi/T)^2 \sin(2t\pi/T_i + \Phi_i)$$

where:
θ_{pi} is peak amplitude
t is time
T_i is period of the movement
Φ_i is the phase between successive terms

Velocity increases as a linear function of frequency and amplitude; acceleration increases as a function of frequency squared and amplitude. According to Marras et al. (1993) the risk of CTDs was significantly

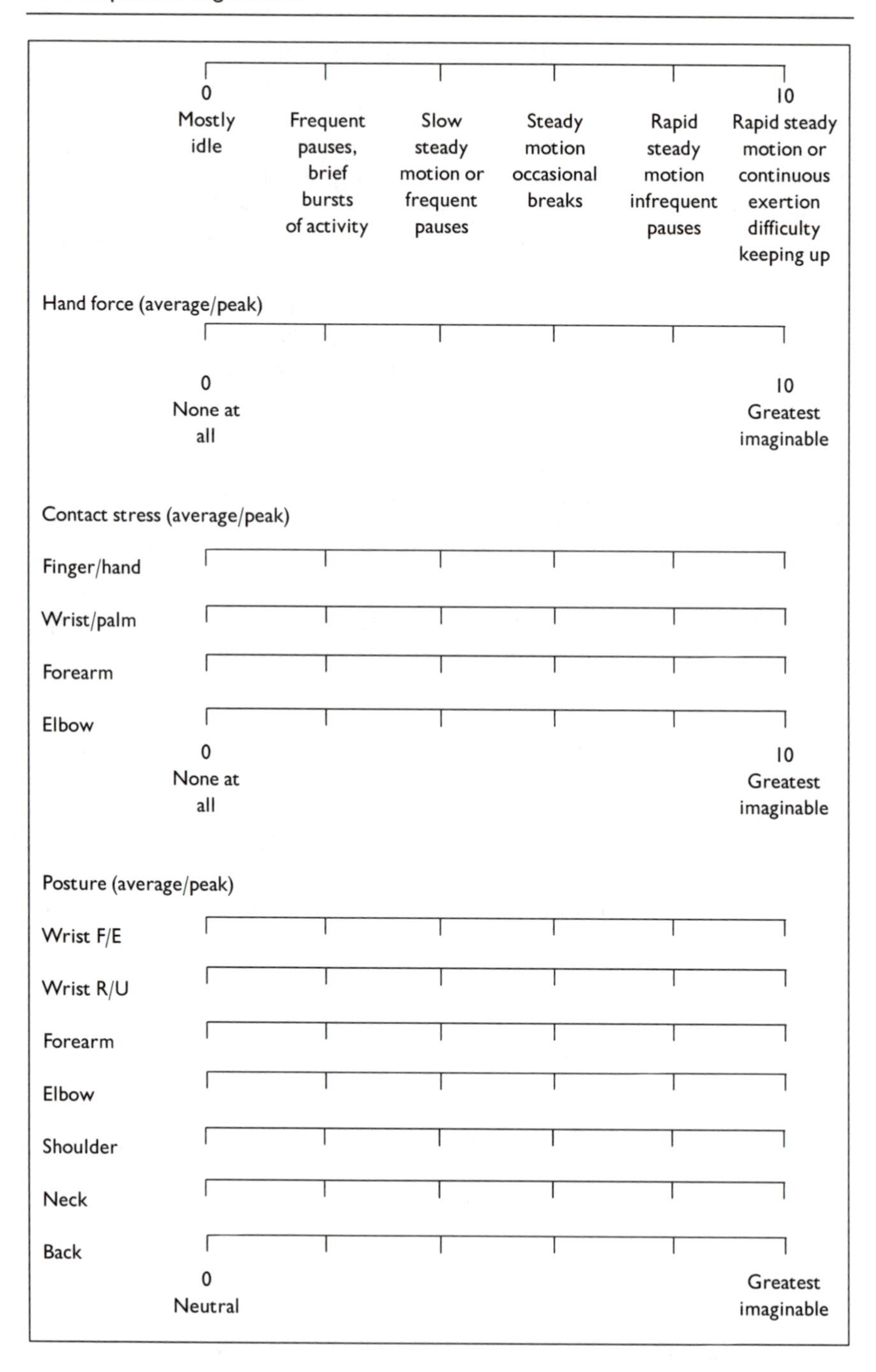

Figure 5.8 Scales rating risk factors of work related musculoskeletal disorders (adapted from Latko et al. 1997).

greater for workers having mean wrist flexion–extension accelerations of 820°/s^2 versus a control group whose mean accelerations were only 490°/s^2.

Electromyography and electromechanical goniometers provide quantitative data about exertions and movement patterns. It is important that the problem be well defined before using EMG and goniometry as they generate large quantities of data. EMG also requires specialized instrumentation and expertise. It is used in the field, but not as a routine survey tool. EMG is generally used in situations where there is a well-defined problem. For example it was used by Gerard et al. (1999) to examine the effect of key stiffness on muscle forearm activity. It can be argued that EMG is a better indicator of muscle force than external finger force.

Force is measured as the external loads on the fingers. Technically force is an array of vectors acting on the palmar side of the fingers. Operationally force is treated as a scalar quantity that corresponds with the exertion effort. Force has units of Newtons, but kilograms and pounds are still widely used. As was stated above, force is often normalized as a percentage of maximum voluntary contraction, or %MVC.

Force can be estimated based on the weight and friction of the work objects and knowledge of the mechanical relationship between the work objects and the hand. For example, holding a suitcase requires the fingers to flex against the weight of the suitcase (see Figure 5.6b). If the suitcase weighs 100 N, then at least 100 N of finger force will be required to pick up and hold the suitcase. Additional force will be required to overcome the inertial forces of the cases. Pulling a 20 N file from a drawer requires a worker to pinch the sides of the file tight enough to keep the file from slipping out of the hand (see Figure 5.6a). The required pinch force will be at least equal to the weight of the file divided by two times the coefficient of friction. Additional force will be required to overcome friction from the adjacent file folders and the inertial force of the file. The physiological processes regulating force of exertion are not perfect and workers will exert more than the minimum required force (Frederick and Armstrong 1987; Westling and Johannson 1984).

Forces can be quantified by workers using rating scales such as that described by Borg (1990). Armstrong, Punnett and Ketner (1989) used a visual analogue scale for workers to rate their perception of hand tool weight where '0 = too light', 5 = just right' and '10 = too heavy'. Nearly all of the tools with mass less than 1.5 kg were rated as 'just right'. While this does not mean that workers will not experience musculoskeletal symptoms for tools with mass less than 1.5 kg, it becomes a good starting point for selection and design of new tools. Snook et al. (1995) reported that one out of 15 subjects performing repetitive hand work for eight hours per day developed carpal tunnel syndrome symptoms within one week even though they were working at frequencies and forces they considered acceptable.

Forces can be quantified by observers using standardized rating scales such as that shown in Figure 5.8. Ratings are from zero to ten where '0 = None' and '10 = Greatest imaginable'. Jobs with force requirements ranging from 'None' to 'Greatest imaginable' should be selected to show how force influences workers behaviour. Pair wise ranking of these jobs can be used to arrange them on the scale. These jobs can be used as references for future force ratings. Low force jobs are characterized by the freedom and fluidity of motion, while high force jobs are characterized by jerky motions and thrusting of the body to apply force. Force ratings should include both peak and average values. Peak values can go up to 10, but it is difficult for workers to sustain exertions with average values greater than 15–20%MVC (Bystrome and Fransson-Hall 1994).

Instrumentation includes use of force sensors under work objects, such as tools and keyboards (Armstrong et al. 1994). Instrumentation for force measurements is widely used in laboratories, but is less widely used in field settings. Use of EMG to evaluate muscle force patterns has already been described (Armstrong et al. 1982). It is particularly useful in jobs where workers hold and use tools, such as knives. Gerard et al. (1999) found that forearm EMG patterns agreed favourably with keyboard reaction force patterns for keyboard work. They also found that both force and EMG activity were related to key stiffness.

Posture refers to the position of one or more joints. Many joints have two or three degrees of freedom, which can make determination of the upper limb posture a tedious process. It is generally desirable to design work systems so that the job can be performed with the elbows at the sides of the body, without extreme forearm rotation, without extreme elbow flexion, without flexing or extending the wrist, and without deviating the wrist side to side. Even the best posture will eventually become uncomfortable and lead to symptoms or other problems. People are dynamic and they need freedom to change postures when they become uncomfortable. The duration for which a given posture can be maintained may vary from one person to the next. Every effort should be made to give individuals control over their posture in much the same way as a driver uses adjustable seats, steering wheel and cruise control to enable them to change positions on long trips in an automobile.

Although subject to individual variations, posture can be anticipated from the information obtained in the job documentation. For example, from the drawing in Figure 5.2 it can be see that the small female will have to flex her wrist or elevate her elbows to reach the keyboard. Even without seeing the worker in Figure 5.2, it can be anticipated that elevation of the shoulder and flexion of the wrist would be required to perform the job.

Postures can be assessed via worker comfort ratings (Armstrong, Punnett and Ketner 1989). It has been shown that work below and above mid-torso height is associated with increased discomfort and elevated risk of musculo-skeleletal symptoms.

Postures can also be observed and rated. Scales for rating peak and average wrist radial/ulnar deviation, wrist flexion/extension, forearm rotation, elbow flexion/extension, elbow elevation, neck flexion, rotation and deviation, and trunk flexion and rotation are shown in Figure 5.8. Ratings are from zero to ten, where 0 is neutral and 10 is the most extreme position possible. All of the joint ratings should be very close to zero when the arm is hanging relaxed at the sides of the body. As with repetition and force, the analyst will want to develop a video library to illustrate the full range of postures for occupations of interest.

As was mentioned above, posture can be measured objectively using electromechanical goniometers. Several units are commercially available that can be attached to a joint to provide an electrical signal proportional to joint position. The signal is generally recorded on the computer and summarized as a frequency histogram and corresponding descriptive statistics. Radwin et al. (1993) have proposed a method for evaluating postures that utilizes frequency analysis to quantify posture recordings from goniometers.

Contact stress refers to mechanical stresses produced on the surface of the skin and passively transferred to underlying tissues (Fransson-Hall 1995; Lundborg 1982). Contact stress has normal components that act perpendicular to the surface and shear stresses components that act parallel to the surface, e.g. friction. This discussion is concerned with the normal stress that may affect underlying nerve and tendon tissues. Normal stress, σ, can be calculated as the normal force, F, divided by the area of contact, A.

$$\sigma = F/A$$

A 10 N force exerted over a one square centimeter area on the pulp of a finger will have a stress concentration of 10^{-3} Pa. Theoretically it is possible to calculate stress concentrations on the surface of the body, but as a practical matter the distribution of stresses is seldom uniform and the peak stresses will be much higher than average values predicted.

Worker and observer ratings are frequently used to identify and rate contact stresses. Using visual analogue scales and simple numerical ratings can increase the objectivity of ratings. Contact stresses and discomfort patterns often vary from person to person. Although, it may be desirable to know what an individual worker feels, it may also be desirable to determine if those feelings are consistent among workers and if widespread tool or process changes are merited. Therefore, multiple workers should be surveyed when possible (Tannen et al. 1986; Karlquist and Björkstén 1990).

Observers can identify contact stresses where workers exert themselves against a tool, part or other work object or where they simply rest on the edge of a machine or work surface. Stresses can then be rated on a scale of zero to ten. Stress concentration ratings are less consistent than repetition or force; however, in the absence of better methods, ratings may be the method of choice. Scales for rating peak and average stresses on the fingers, the wrist and palm, the forearm and the elbow are shown in Figure 5.8. Analysts should calibrate their ratings by finding jobs that exemplify the extremes and intermediate values. No contact stress on the wrist corresponds to wrist hanging at the side of the body. Maximum contact stress imaginable might correspond to someone pounding with his or her wrist or pressing with his or her fingers against the sharp edges of a connector.

Low temperature and vibration can be identified by observations and then measured using appropriate instrumentation. Vibration is discussed in other chapters of this book.

5.3 DESIGN

Job design can be characterized as a systematic process of determining the job attributes required to achieve a given objective. The previous section was concerned with describing job objectives and attributes. It suffices to say that productivity, quality and general safety are important objectives. This section focuses on the objective of controlling in so far as possible the risk of work related musculoskeletal disorders through the control of exposures to the physical risk factors described above.

5.3.1 The Job Design Process

The job design process is summarized in Figure 5.9. Following the initial analysis of the job, the designer must determine alternative processes, equipment or methods for controlling the risk factors. Alternative designs can be evaluated in terms of four criteria. The first criteria are elimination of the risk factors. In some cases it may not be economically feasible to eliminate the stress. For example, it may not be possible to reduce repetition without excessively reducing production. In other cases it may be possible to completely eliminate a posture stress simply by re-positioning the work. The second criteria are comparison with findings from biomechanical, physiological, psychophysical and epidemiological studies (ACGIH 2000; Bystrome and Fransson-Hall 1994; Snook et al. 1995). The third criteria are successful placement of an injured worker based on feedback from workers and supervising health care professionals. The fourth criteria are reduced incidence or severity of WMSDs with respect to background levels (see Chapter 3). As a practical matter, small populations, worker turnover and job changes and production changes make such comparisons difficult if not impossible. The before and after analysis of risk factors is most frequently used by practitioners.

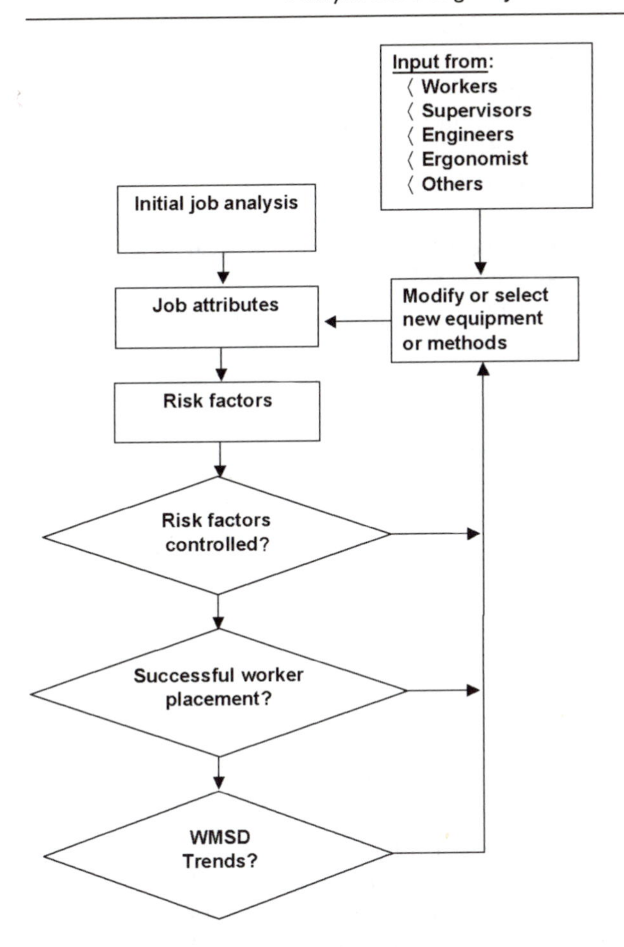

Figure 5.9 The design processes begin with a basic analysis of the job to specify job attributes and ergonomic stresses. Team input is then used to identify selected processes, equipment and methods that control risk factors, result in a successful job placement or control WMSD trends.

5.3.2 Before and After Analysis of Risk Factors

The use of before and after job analysis is shown with two jobs, one from a manufacturing setting and the other from an office setting. These examples are hybrids and were synthesized from real cases to illustrate the design process.

Example 1: Carton packing

Table 5.4 shows a carton packing job in which the worker performs 1,500 exertions per hour with each hand to get and erect cartons, get and pack parts, close and tape flaps, and transfer the filled carton to a pallet. It can be seen that carton packing exposes the worker to high repetition and high peak wrist contact stress and medium peak force, wrist contact stress, peak wrist flexion and extension and radial/ulnar deviation. A conveyor and automatic taping machine were introduced and reduced the required exertions by 20%. Because one of the eliminated exertions, transfer filled carton to pallet was longer than the other exertions, it was predicted that the overall repetition level would be reduced 25%. It was also predicted that the peak force would be reduced from 6 to 3. Wrist contact stress and posture stress are mainly associated with erecting and filling the cartons, so they are not likely to be affected.

Table 5.4 Ergonomic stresses associated with packing job were rated on a scale of 0–P as described by Latko et al. (1997).

Work related risk factor	*Before intervention*	*After intervention‡*
Repetition*	8	6
Force†	6/1.5	3/1
Wrist contact stress†	5/2	5/2
Wrist flexion/extension	5/3	5/3
Wrist radial/ulnar deviation	5/3	5/3

* Average.
† Peak/average.
‡ Intervention: conveyor to take away filled carton and close flaps.

There is increasing literature by which the acceptability of the ergonomic stresses can be evaluated. For example, while the risk of WMSDs is reduced by the intervention, it would still be elevated (Latko et al. 1999). According the ACGIH, we find that the combination of repetition and force is below the proposed TLV, but above the proposed action limit. The combination of medium wrist contact stress, wrist flexion/extension and wrist radial/ulnar deviation supports lowering the acceptable action limit and TLV.

Example 2: Medical Transcription

In Example 2 a transcriptionist types 100 words per minute to transfer medical dictation from a tape recording to a computer via a keyboard. This job and workstation correspond to the drawing shown in Figure 5.2a. This job entails exposure to high repetition, high peak and average wrist flexion, high peak wrist deviation and medium wrist contact stress. The proposed intervention consisted of a new keyboard and an adjustable keyboard tray. The new keyboard required only a light touch (<0.4 N) and could be adjusted around its vertical axis to reduce ulnar deviation. A second intervention was the reprogramming of the software so that the worker did not have to use the control keys in the lower right and left hand corners of the keyboard in combination with other keys to execute program commands. The commands were reassigned to keys at the top of the keyboard. Finally a keyboard tray was retrofitted to the workstation so that the worker could position the keyboard at elbow height (see Figure 5.2b).

Table 5.5 Ergonomic stresses association with medical transcript (see Figure 5.8a) were rated on a scale of 0–10 as described by Latko et al. 1997.

Work related risk factor	*Before intervention*	*After intervention‡*
Repetition*	9	9
Force†	1.5/1	3/1
Wrist contact stress†	6/5	2/1
Wrist flexion/extension†	7/6	3/2
Wrist radial/ulnar deviation†	7/2	2/1

* Average.
† Peak/average.
‡ Interventions (see Figure 1b).

- New keyboard with light touch keys to reduce peak force and variable geometry to reduce ulnar deviation while keying.
- Software re-programmed to use function keys across keyboard in place of Ctrl key in lower right and left-hand corners of keyboard.
- Keyboard tray used to adjust keyboard to elbow height.

5.3.3 Other Design Considerations

In addition to reducing WMSD risk factors, the job should be examined to make sure that the worker can still perform the job and meet the design objectives and original production standards. In addition to the basic time and motion study techniques described above, the reader is referred to methods for assessing usability described by Rubin (1994).

The details of the job design require specific knowledge of the process and go beyond the scope of this discussion. It suffices to say that the process can be facilitated by involvement of workers and others with site specific and general knowledge of the process. A team approach is commonly advocated for controlling WMSD risk factors (McKenzie et al. 1985; Mansfield and Armstrong1997; Moore et al. 1990). The team should be convened as early in the design process as possible when the cost of making changes is least. Unfortunately, the ergonomist and the team all too often become involved after a job enters production. Making changes other than minor adjustments at this stage can be cost prohibitive. One of the reasons ergonomic teams are not convened in the design process is that the jobs don't yet exist; however, it may be possible to find and analyse similar jobs.

Finally, job designers need to recognize that there are important physical and behavioural differences in the worker population and that the ideal job design is not the same for all workers. The preferred work height will vary from person to person of the same size. Workers will exert varying degrees of force on the same keyboard. This does not mean that one worker is doing the job wrong – it only means that they do it differently. It is important that job designers recognize these differences and accommodate them in the same way that vehicle designers accommodate them with adjustable seats, steering wheels, arm rests and mirrors. Fortunately there is an abundance of anthropometric data available. Unfortunately the available data do not capture the behavioural differences between workers and fitting trials are often required to determine specifications for a particular application (Pheasant 1996; Ulin et al. 1992). Computerized biomechanical models are being developed and show great promise for capturing these variations for certain tasks (Chapter 13).

5.4 SUMMARY

This chapter is concerned with analysis and design of job for both primary and secondary prevention of disability associated with WMSDs. Job analysis is a systematic process for obtaining information about a job. In this case the desired information is concerned with what the worker does and how that exposes him or her to WMSD risk factors. Information needed to characterize what the worker does, sources of information for determining what the worker does, the relationship between what the worker does and his or her exposure to physical risk factors of WMSDs and finally recognition and quantification of risk factors have been described.

Job design can be characterized as a systematic process of determining the job attributes required to achieve a given objective. Following the initial analysis of the job, the designer must determine alternative processes, equipment or methods for controlling the risk factors. Job design alternatives can be evaluated by: elimination of the risk factors; comparisons with findings from biomechanical, physiological, psychophysical and epidemiological studies; the successful placement of injured workers; and reductions of incidence or severity of WMSDs with respect to background levels. Evaluation of jobs should include consideration of usability by workers. Designs that work well for one worker may not work well for another. Designers must consider behavioural as well as physical differences among workers. Job design is facilitated by a team approach that includes worker involvement.

5.5 REFERENCES

ACGIH. TLVs®, BEIs®, *Threshold Limit Values for Chemical Substances and Physical Agents and Biological Exposures Indices*. Cincinnati, OH, 2000: 119–121.

Armstrong T, Chaffin D. Some biomechanical aspects of the carpal tunnel. *J Biomech* 1979; 12(7), 567–570.

Armstrong T, Foulke J, Joseph B, Goldstein S. Investigation of cumulative trauma disorders in a poultry processing plant. *Am Ind Hyg Assoc J* 1982; 43(2), 103–116.

Armstrong T, Castelli W, Evans F, Diaz-Perez R. Some histological changes in carpal tunnel contents and their biomechanical implications. *J Occ Med* 1984; 26(3), 197–201.

Armstrong T, Punnett L, Ketner P. Subjective worker assessments of hand tools used in automobile assembly. *Am Indus Hyg Ass J* 1989; 50(12), 639–645.

Armstrong T, Buckle P, Fine L, Hagberg M, Jonsson B, Kilbom A, Kourinka I, Silverstein B, Sjogaard G, Viikari-Juntura E. A conceptual model for work-related neck and upper limb disorders. *Scand J Work Environ Health* 1993; 19(2), 74–84.

Armstrong T, Foulke J, Martin B, Gerson J, Rempel D. Investigation of applied forces in alphanumeric keyboard work. *Am Ind Hyg Assoc J* 1994; 55(1), 30–35.

Barnes R. *Motion and Time Study, Design and Measurement of Work*. John Wiley and Sons, New York, 1980.

Birt J, Snyder M, Duncanson J. Appendix D 'Standard' verbs. In *Human Factors Design Guide for Acquisition of Commercial-off-the-shelf Subsystems, Non-Developmental Systems*, DOT/FAA/CT-96/1. National Technical Information Service, Springfield, VA, 1996.

Bobjer O, Johansson S, Piguet S. Friction between hand and handle. Effects of oil and lard on textured and non-textured surfaces; perception of discomfort. *Appl. Ergonomics* 1993; 24(3), 190–202.

Bongers P, de Winter C, Kompier M, Hildebrandt V. Psychosocial factors at work and musculoskeletal disease. *Scand J Work Environ Health* 1993; 19, 297–312.

Borg G. Psychophysical scaling with applications in physical work and the perception of exertion. *Scand J Work Environ Health* 1990; 16(Suppl 1), 55–58.

Buchholz B, Frederick L, Armstrong T. An investigation of human palmar skin friction and the effects of materials, pinch force and moisture. *Ergonomics* 1988; 31(3), 317–325.

Bystrom S, Fransson-Hall C. Acceptability of intermittent handgrip contractions based on physiological response. *Human Factors* 1994; 36(1), 158–171.

Niebel B, Freivalds A. *Methods Standards and Work Design*, 10th edition McGraw-Hill, New York, 1999.

Frederick L, Armstrong T. An investigation of friction and weight on pinch force. *Ergonomics* 1995; 38(12), 2447–2454.

Gerard M, Armstrong T, Franzblau A, Martin B, Rempel D. The effects of keyboard stiffness on typing force, finger electromyography, and subjective discomfort. *Am Indust Hyg Assoc J* 1999; 60(6), 762–769.

Gilbreth F, Gilbreth L. *Motion Study for Crippled Soldiers*. A paper presented at a meeting of the American Association for the Advancement of Science, in Columbus, OH, 27 December 1915, pp. 281–288. In *The Writings of the Gilbreths* W Spriegel, C Myers, eds. Richard D. Irwin, Inc., Homewood, IL, 1953.

Goldstein S, Armstrong T, Chaffin D, Matthews L. Analysis of cumulative strain in tendons and tendon sheaths. *J Biomech* 1987; 20: 1–6.

Homan M. *Evaluation of exposure assessment and diagnostic methodologies for use in epidemiological studies of work-related musculoskeletal disorders.* Ph.D. Dissertation, University of Michigan, Ann Arbor, MI 48109, 1999.

Jonsson B. The static load component in muscle work. *Eur J Appl Phys* 1988; 57, 305–310.

Karlquist L, Björkstén M. Design for prevention of work related musculskeletal disorders. In *Ergonomics: The Physiotherapist in the Workplace,* M Bullock, ed. Churchill Livingstone, New York, 1990: 155–157.

Kilbom A. Repetitive work of the upper extremity: Part I – Guidelines for the practitioner. *Int J Indust Ergonomics* 1994a; 14, 51–57.

Kilbom A. Repetitive work of the upper extremity: Part II – The scientific basis (knowledge base) for the guide. *Int J Indust Ergonomics* 1994b; 14: 59–86.

Latko W, Armstrong T, Franzblau A, Ulin S, Werner R, Albers J. A cross-sectional study of the relationship between repetitive work and the prevalence of upper limb musculoskeletal disorders. *Am J Indust Med* 1999; 36: 248–259.

Latko W, Armstrong T, Foulke J, Herrin G, Rabourn R, Ulin S. Development and evaluation of an observational method for assessing repetition in hand tasks. *Am Indust Hyg Ass J* 1997; 58(4), 278–285.

Lundborg G, Gelberman R, Minteer-Convery M, Lee Y, Hargens A. Median nerve compression in the carpal tunnel – Functional response to experimentally induced controlled pressure. *J Hand Surg* 1982; 7(3), 252–259.

Marras W, Schonmarklin R. Wrist motions in industry. *Ergonomics* 1993; 36(4), 341–351.

Mansfield J, Armstrong T. Library of Congress workplace ergonomics program. *Am Ind Hyg Assoc J* 1997; 58, 138–144.

Mathiassen S, Winkel J. Quantifying variation in physical load using exposure-vs-time data. *Ergonomics* 1991; 34(12), 1455–1468.

McAtamney L, Corlett E. RULA: a survey method for the investigation of work-related upper limb disorders. *Appl Ergonomics* 1993; 24(2), 91–99.

McCormick E. Job and task analysis. Chapter 2.4 in *Handbook of Industrial Engineering* G Salvendy, ed. John Wiley, New York, 1982: 244–246.

McKenzie F, Storment J, Van Hook P, Armstrong T. A program for control of cumulative trauma disorders in an electronics plant. *Am Ind Hyg Assoc J* 1985; 46(11), 674–678.

Moore S. Flywheel truing — a case study of an ergonomic intervention. *AIHA J* 1994; 55(3), 236–244.

Moore S, Garg A. The strain index: A proposed method to analyze jobs for risk of distal upper extremity disorders. *AIHA J* 1995; 56(5), 443–458.

Niebel B, Freivalds A. *Methods, Standards and Work Design*, 3rd edition. McGraw-Hill, Boston, 1999.

Phalen G. The carpal-tunnel syndrome. *J Bone Joint Surg* 1966; 48A: 211–228.

Pheasant S. Body *Space: Anthropometry, Ergonomics and the Design of Work*, Taylor & Francis Ltd, London, 1996.

Radwin R, Lin M. An analytical method for characterizing repetitive motion and postural stress using spectral analysis. *Ergonomics* 1993; 36(4), 379–389.

Rempel D, Bach J, Gordon L, So Y. Effects of forearm pronation/supination on carpal tunnel pressure. *J Hand Surg* 1998; 23A: 38–42.

Rempel D, Keir P, Smutz W, Hargens A. Effects of static fingertip loading on carpal tunnel pressure. *J Ortho Res* 1997; 15: 422–426.

Rubin J. *Handbook of Usability Testing: How to Plan, Design, and Conduct Effective Tests*, John Wiley & Sons, Inc., New York, 1994.

Sinclair M. Subjective assessment. Chapter 3 in *Evaluation of Human Work. A Practical Methodology,* second edition, J Wilson, N Corlett, eds. Taylor & Francis, London, 1995: 69–100.

Snook S, Vaillancourt D, Ciriello VM, Webster B. Psychophysical studies of repetitive wrist flexion and extension. *Ergonomics* 1995; 38(7): 1488–1507.

Tannen K, Stetson D, Silverstein B, Fine L, Armstrong T. An evaluation of scissors for control of upper extremity disorders in an automobile upholstery plant. In Karwoski W (ed.), *Trends in Ergonomics/Human Factors III*. North-Holland, New York, NY, 1986: 631–639.

Ulin S, Snook S, Armstrong T, Herrin G. Preferred tool shapes for various horizontal and vertical work locations. *Appl Occup Environ Hyg* 1992; 7(5): 327–337.

Werner R, Armstrong T, Bir C, Aylard M. Intra carpal canal pressures: The role of finger, hand, wrist and forearm position. *Clin Biomech (England)* 1997; 12(1): 44–51.

Westling G, Johansson R. Factors influencing the force control during precision grip. *Exp Brain Res* 1984; 53: 277–284.

Zandin K, Gardner D, Gill E, Wilk J. ErgoMOST: An engineer's tool for measuring ergonomic stress. In *Occupational Ergonomics: Theory and Applications,* A Bhattacharya, J McGlothlin, eds. Marcel Dekker, Inc., New York, 1996: 417–429.

Chapter 6

The Psychophysical Approach to Risk Assessment in Work-Related Musculoskeletal Disorders

Stover H. Snook

6.1 HISTORY OF PSYCHOPHYSICS

Psychophysics is a very old branch of psychology that is concerned with the relationship between physical stimuli that occur in the 'outside world', and the sensations they produce in the body's 'inside world'. It began 164 years ago with the investigations of Ernst Heinrich Weber (1795–1878), a professor of anatomy (and of physiology) at the University of Leipzig in the early 1800s (Boring 1950; Woodworth and Schlosberg 1954).

Weber was very interested in the sense of touch. In 1834 he ran some experiments on the perception of weight, and found that weight must be increased by a constant fraction of its value to be just noticeably different (JND). This constant fraction is about 1/40, and is independent of the magnitude of weight. In other words, if there is a 40 gram weight in one hand, there must be a 41 gram weight in the other hand to notice a difference between the two weights. If there is an 80 gram weight in one hand, there must be an 82 gram weight in the other hand to notice a difference. This became known as Weber's Law, $\Delta I/I = \text{constant}$, where I is the weight (intensity) and ΔI is the increment that the weight must be increased to be just noticeably different.

Another professor at the University of Leipzig became very interested in Weber's research. Gustav Theodor Fechner (1801–1887) was a professor of physics, and active in studying the relationship between body and mind (Boring 1950; Woodworth and Schlosberg 1954). Fechner expanded upon Weber's work and, in 1860, published a book called *Elemente der Psychophysik*, in which he proposed what is now known as Fechner's Law. This law states that the strength of a sensation (S) is directly related to the intensity of its physical stimulus (I) by means of a logarithmic function ($S = k \log I$). The constant k is a function of the particular unit of measurement used.

Fechner's Law was known to be quite accurate in the middle ranges of stimuli and sensations, but not so accurate at the extremes. In 1950 S. S. Stevens, a professor of psychology at Harvard University, proposed that the relationship between stimuli and sensations was not a logarithmic one, but a power function ($S = kI^n$) (Stevens 1960). Stevens and his colleagues collected data to show that the power function was accurate throughout the entire range of stimuli and sensations.

When plotted in log–log coordinates, a power function is represented by a straight line, with the exponent (n) being equal to the slope of the line. Over the years, exponents have been experimentally determined for many types of stimuli. For example, 3.5 is the exponent for electric shock, 1.3 for taste (salt), and 0.6 for loudness (binaural). Of particular interest in this paper is the perception of muscular effort and force, both of which have been found to obey the power law, and both with an exponent of approximately 1.6 (Borg 1962; Eisler 1962).

6.2 APPLICATION OF PSYCHOPHYSICS

Psychophysics has been applied to practical problems in many areas. For example, psychophysics has been used to develop the scales of loudness (decibel scale), effective temperature (EF scale), and brightness (bril scale) (Stevens, 1960, 1956; Houghton and Yagloglow 1923). Psychophysics has also been used by Borg (1962, 1973) in developing ratings of perceived exertion (RPE), and by Caldwell et al. (1973, 1967) in the development of effort scales. In addition, psychophysics has been utilized in the design of stairways (Irvine et al. 1990), chairs (Coleman et al. 1998) and VDT workstations (Grandjean 1983).

6.2.1 Manual Handling Tasks

The first applications of psychophysics to manual handling tasks were two U.S. Air Force studies of lifting. The Air Force studies used young, male college students to investigate the loading of ammunition cases into F-86H aircraft, as well as other operational and maintenance activities. Unfortunately, the researchers did not control for repetition rate, training or fitness; nor did they mention or discuss the concept of psychophysics (Emanuel et al. 1956; Switzer 1962).

About 30 years ago, the Liberty Mutual Research Center began to apply psychophysics to manual handling tasks, expanding upon the earlier Air Force studies (Snook and Irvine 1967). The purpose was to develop recommendations for use in reducing industrial low back compensation claims. At that time, psychophysics was the only method that could yield usable data for evaluating manual handling tasks. In the Liberty Mutual studies, the worker was given control of one of the task variables, usually the weight of the object being handled (Snook 1978; Snook and Ciriello 1991). All other variables such as repetition rate, size, height, distance, etc., were controlled. The worker then monitored his or her own feelings of exertion or fatigue, and adjusted the weight of the object accordingly. It was believed that only the individual worker could sense the various strains associated with manual handling tasks; and only the individual worker could integrate the sensory inputs into one meaningful response.

Eleven separate experiments were conducted at the Liberty Mutual Research Center over a period of 25 years. Each experiment lasted two to three years. These experiments were unique in that they used realistic manual handling tasks performed by industrial workers (male and female) over long periods of time (at least 80 hours of testing for each subject). Teams of three workers performed lifting, lowering, pushing, pulling, and carrying tasks. Measurements of oxygen consumption and heart rate were recorded for comparison with psychophysical measurements. The experimental design also included 16 to 20 hours of training and conditioning, and a battery of 41 anthropometric measurements from each subject. The results of these experiments were combined and integrated into tables of maximum acceptable weights and forces for various percentages of the population (Snook 1978; Snook and Ciriello 1991).

6.2.2 Repetitive Motion Tasks

The success of the psychophysical approach with manual handling tasks led to the application of this approach to the repetitive motion tasks of the hands and wrists. The purpose of these studies was to develop recommendations for use in reducing the rising number of compensation claims for cumulative trauma or repetitive motion disorders. However, the application of psychophysics to repetitive motion tasks was more difficult to accomplish. For example, the nature of the cumulative trauma disorder was different from low back pain, and more hours of testing were required from each subject.

Four psychophysical cumulative trauma experiments were conducted using the same basic experimental design. Teams of three to eight female workers performed repetitive motion tasks for seven hours per day, five days per week, for four to five weeks. The tasks included wrist flexion and extension (with both power and pinch grips), ulnar deviation, and handgrip. In an effort to provide maximum protection for the subjects, symptoms were recorded every hour of the experiment. Each subject was asked to record any feelings of soreness, stiffness, or numbness from the fingers and thumb, hand, and forearm. Vibrometry testing was also used to measure changes in the tactile sensitivity of the finger-tips.

The maximum acceptable forces for various repetition rates of wrist flexion, extension, and ulnar deviation are presented in Table 6.1. The preliminary results from the handgrip motion are also presented in Table 6.1. Repetition rates of 10, 15, and 20/min were used for flexion (power and pinch grips) and extension (power grip). Installation of a faster computer allowed faster repetition rates (15, 20, and 25/min) for extension (pinch grip), ulnar deviation, and handgrip. The maximum acceptable forces are given for different percentages of the population. It is recommended that tasks should not exceed forces for 75% of the population, and ideally, 90% of the population.

Table 6.1 Maximum acceptable forces for repetitive hand and wrist motions (females) (Newtons).

Motion	*Grip*	*Percent of population*	*Repetition rate* 10/min	15/min	20/min	25/min
Flexion (Snook et al. 1995)	Power N=14	90	13.5	12.0	10.2	
		75	20.9	18.6	15.8	
		50	29.0	26.0	22.1	
		25	37.2	33.5	28.4	
		10	44.6	40.1	34.0	
	Pinch N=15	90	7.4	7.4	6.0	
		75	11.5	11.5	9.3	
		50	16.0	16.0	12.9	
		25	20.6	20.6	16.6	
		10	24.6	24.6	19.8	
Extension (Snook et al. 1995, 1999)	Power N=15	90	7.8	6.9	5.4	
		75	12.1	10.9	8.5	
		50	16.8	15.1	11.9	
		25	21.5	19.3	15.2	
		10	25.8	23.2	18.3	
	Pinch N=20	90		2.7	1.9	1.0
		75		5.1	4.5	3.4
		50		7.8	7.4	6.0
		25		10.5	10.3	8.6
		10		12.9	12.9	11.0
Ulnar Deviation (Snook et al. 1997)	Power N=24	90		4.5	4.5	4.3
		75		9.0	8.9	8.7
		50		14.0	13.9	13.5
		25		19.0	18.9	18.3
		10		23.5	23.3	22.6
Handgrip (Snook 1999)	Power N=14	90		13.6	12.2	10.0
		75		24.5	22.0	18.1
		50		36.6	32.8	27.0
		25		48.7	43.6	35.9
		10		59.6	53.4	44.0

6.3 CRITIQUE OF PSYCHOPHYSICS

Every methodology has its advantages and disadvantages, and psychophysics is no exception. The major advantages and disadvantages of psychophysics have been reviewed by Snook (1985) and Ayoub and Dempsey (1999). The advantages include the following:

- Psychophysics permits the realistic simulation of industrial work.
- Psychophysics can be used to study the very intermittent manual handling tasks, and the very fast repetitive motion tasks.
- Psychophysical results are consistent with the industrial engineering concept of a 'fair day's work for a fair day's pay'.
- Psychophysical results are very reproducible.
- Psychophysics can be used to measure subjective variables such as pain, fatigue, and discomfort – variables that cannot be measured objectively.
- The psychophysical approach is less costly and time consuming to apply in industry than many of the biomechanical and physiological techniques.
- Psychophysics is particularly useful in exposing subjects to hazardous tasks without excessive risk.

The primary disadvantages of psychophysics include the following:

- Psychophysics is a subjective method that relies on self-report from subjects. It will probably be replaced when and if more objective methods become available.
- Psychophysical values from very fast frequency manual handling tasks are higher than recommended metabolic criteria.
- Psychophysics does not appear sensitive to the bending and twisting motions that are often associated with the onset of low back pain.

Psychophysics is a particularly useful and versatile tool for evaluating jobs. Psychophysics is best used in conjunction with biomechanical, physiological, and epidemiological methods, e.g., the development of the NIOSH lifting equation in the U.S.A. (Waters et al. 1993). Several studies have concluded that recommendations based upon psychophysical results can reduce low back disorders in industry (Snook et al. 1978; Liles et al. 1984; Herrin et al. 1986; Marras et al. 1999). However, there have been no studies validating the effect of psychophysics on cumulative trauma disorders of the hands and wrist. There is a critical need for further validation studies to be performed.

6.4 REFERENCES

Ayoub MM, Dempsey PG. The psychophysical approach to manual materials handling task design. *Ergonomics* 1999; 42(1), 17–31.

Borg GAV. *Physical Performance and Perceived Exertion* Munksgaard, Copenhagen, 1962.

Borg GAV. Perceived exertion: A note on 'history' and methods. *Medicine and Science in Sports* 1973; 5, 90–93.

Boring EG. *A History of Experimental Psychology*. Appleton-Century-Crofts, New York, 1950.

Caldwell LS, Grossman EE. Effort scaling of isometric muscle contractions. *J Motor Behavior* 1973; 5, 9–16.

Caldwell LS, Smith RP. *Subjective Estimation of Effort, Reserve, and Ischemic Pain*. US Army Medical Research Labs, Report No. 730, Fort Knox, Kentucky, 1967.

Coleman N, Hull BP, Ellitt G. An empirical study of preferred settings for lumbar support on adjustable office chairs. *Ergonomics* 1998; 41, 401–419.

Dempsey PG. A critical review of biomechanical, epidemiological, physiological and psychophysical criteria for designing manual materials handling tasks. *Ergonomics* 1998; 42, 73—88.

Eisler H. Subjective scale of force for a large muscle group. *J Experimental Psychology* 1962; 64, 253–257.

Emanuel I, Chaffee JW, Wing J. *A Study of Human Weight Lifting Capabilities for Loading Ammunition into the F-86H Aircraft*. Wright Air Development Center, WADC-TR-56-367, Wright-Patterson Air Force Base, Ohio, 1956.

Grandjean E, Hünting W, Pidermann M. VDT workstation design: preferred settings and their effects. *Human Factors* 1983; 25, 161–175.

Herrin GD, Jaraiedi M, Anderson CK. Prediction of overexertion injuries using biomechanical and psychophysical models. *Am Industrial Hygiene Assoc J* 1986; 47, 322–330.

Houghton FC, Yagloglou CP. Determination of the comfort zone. *J Am Soc Heating Ventilating Engineers* 1923; 29, 515–536.

Irvine CH, Snook SH, Sparshatt JH. Stairway risers and treads: Acceptable and preferred dimensions. *Applied Ergonomics* 1990; 21, 215–225.

Liles DH, Deivanayagam S, Ayoub MM, Mahajan P. A job severity index for the evaluation and control of lifting injury. *Human Factors* 1884; 26, 683–693.

Marras WS. The effectiveness of commonly used lifting assessment methods to identify industrial jobs associated with elevated risk of low-back disorders. *Ergonomics* 1999; 42, 229–245.

Nicholson AS. A comparative study of methods for establishing load handling capabilities. *Ergonomics* 1989; 32, 1125–1144.

Snook SH. The design of manual handling tasks. *Ergonomics* 1978; 21, 963–985.

Snook SH. Psychophysical considerations in permissible loads. *Ergonomics* 1985; 28, 327–330.

Snook SH. Unpublished data, 1999.

Snook SH. Campanelli RA, Hart JW, A study of three preventive approaches to low back injury. *J Occupational Medicine* 1978; 20, 478–481.

Snook SH, Ciriello VM. The design of manual handling tasks: revised tables of maximum acceptable weights and forces. *Ergonomics* 1991; 34, 1197–1213.

Snook SH, Ciriello VM, Webster BS. Maximum acceptable forces for repetitive wrist extension with a pinch grip. *Int J Industrial Ergonomics* 1999; 24, 579–590.

Snook SH, Irvine CH. Maximum acceptable weight of lift. *Am Industrial Hygiene Assoc J* 1967; 28, 322–329.

Snook SH, Vaillancourt DR, Ciriello VM, Webster BS. Psychophysical studies of repetitive wrist flexion and extension. *Ergonomics* 1995; 38, 1488–1507.

Snook SH, Vaillancourt DR, Ciriello VM, Webster BS. Maximum acceptable forces for repetitive ulnar deviation of the wrist. *Am Industrial Hygiene Assoc J* 1997; 58, 509–517.

Stevens SS. The direct estimation of sensory magnitudes loudness. *Am J Psychology* 1956; 69, 1–25.

Stevens SS. The psychophysics of sensory function. *Am Scientist* 1960; 48, 226–253.

Switzer SA. Weight lifting capabilities of a selected sample of human males. Aerospace Medical Research Labs, MRL-TDR-62-57. Wright-Patterson Air Force Base, Ohio, 1962.

Waters TR, Putz-Anderson V, Garg A, Fine LJ. Revised NIOSH equation for the design and evaluation of manual lifting tasks. *Ergonomics* 1993; 36, 749–776.

Woodworth RS, Schlosberg H. *Experimental Psychology*, revised edition. Henry Holt, New York, 1954.

Chapter 7

Health Disorders Caused by Occupational Exposure to Vibration

Massimo Bovenzi

7.1 INTRODUCTION

Mechanical vibration arises from a wide variety of processes and operations performed in industry, mining and construction, forestry and agriculture, and public utilities. ***Whole-body vibration*** occurs when the human body is supported on a surface which is vibrating, e.g. in all forms of transport and when working near some industrial machinery. ***Hand-transmitted vibration*** occurs when the vibration enters the body through the hands, e.g. in various work processes where rotating or percussive power tools or vibrating workpieces are held by the hands or fingers.

The human response to vibration depends mainly on the magnitude, frequency and direction of the vibration signal (Griffin 1990). The magnitude of vibration is quantified by its displacement (m), its velocity ($m\,s^{-1}$), or its acceleration ($m\,s^{-2}$). For practical convenience, the magnitude of vibration is expressed in terms of an average measure of the acceleration, usually the root mean square value ($m\,s^{-2}$ r.m.s.). The r.m.s. magnitude is related to the vibration energy and hence the vibration injury potential. The frequency of vibration is expressed in cycles per second and it is measured in Hertz (Hz). Biodynamic investigations have shown that the response of the human body to vibration is frequency dependent (Griffin 1990). The adverse health effects of whole-body vibration can occur in the low frequency range from 0.5 to 80 Hz. For hand-transmitted vibration, frequencies from 6.3 to 1250 Hz can provoke disorders in the hand–arm system. Frequencies below about 0.5 Hz can cause motion sickness. To account for the differences in the response of the body to vibration frequency, current standards for human vibration recommend to weight the frequencies of the measured vibration according to the possible deleterious effect associated with each frequency (ISO 5349:1986; ISO 2631-1 : 1997). Frequency weightings are required for three orthogonal directions (x-, y- and z-axes) at the interfaces between the body and the vibration. The vibration total value (a_v) of the weighted r.m.s. acceleration (also known as the vector sum) is then calculated as the root-sum-of-squares of the frequency weighted acceleration values (a_w) for the three measured axes of vibration in ms^{-2}. For the health effects of whole-body vibration on seated persons (ISO 2631-1 : 1997): $a_v = (k_x^2 a_{wx}^2 + k_y^2 a_{wy}^2 + k_z^2 a_{wz}^2)^{1/2}$, where $k = 1.4$ for x- and y-axes and $k = 1$ for z-axis. For the health effects of hand-transmitted vibration on the upper limbs of the exposed workers (ISO 5349 : 1986): $a_{hv} = (a_{hwx}^2 + a_{hwy}^2 + a_{hwz}^2)^{1/2}$.

In addition to the physical characteristics of vibration, some other factors are believed to be related to the injurious effects of vibration, such as the duration ofexposure (daily, yearly and lifetime cumulative exposures), the pattern of exposure (continuous, intermittent, rest periods), the type of tools, processes or vehicles which produce vibration, the environmental conditions (ambient temperature, airflow, humidity, noise), the dynamic response of the human body (mechanical impedance, vibration transmissibility, absorbed energy), and the individual characteristics (method of tool handling or style of vehicle driving, body posture, health status, training, skill, use of personal protective equipment, individual susceptibility to injury).

Exposure to harmful vibration at the workplace can induce several complaints and health disorders, mainly at the upper limbs and the lower back. This study summarizes the long-term effects caused by occupational exposure to whole-body vibration and hand-transmitted vibration.

7.2 WHOLE-BODY VIBRATION

7.2.1 General

Long-term occupational exposure to intense whole-body vibration (WBV) is associated with an increased risk for disorders of the lumbar spine and the connected nervous system (CR 12349:1996). With a lower probability, the neck/shoulder, the gastrointestinal system, the female reproductive organs, the peripheral veins, and the cochleo-vestibular system are also assumed to be affected by WBV (CR 12349:1996). However, there is a weak epidemiologic support for WBV induced disorders of organ systems other than the lower back. Epidemiologic studies have pointed out that drivers of off-road vehicles, industrial vehicles and machines, buses, and helicopter pilots are occupational groups at greater risk for low back disorders than other worker groups unexposed to WBV (Bongers and Boshuizen 1990). It has been estimated that 4–7% of all employees in the U.S., Canada and some European countries are exposed to potentially harmful WBV (CR 12349:1996). In some countries (e.g. Belgium, France, Germany, The Netherlands) back disorders occurring in workers exposed to WBV are, under certain conditions regarding intensity and duration of exposure, considered to be an occupational disease which is compensable.

The role of WBV in the etiopathogenesis of low back injuries is not yet fully clarified as driving of vehicles involves not only exposure to harmful

WBV but also to several ergonomic factors which can affect the spinal system, such as prolonged sitting in a constrained posture, bending forward and frequent twisting of the spine. Moreover, some driving occupations involve heavy lifting and manual handling activities (e.g. drivers of delivery trucks), which are known to strain the lower part of the back. Individual characteristics (age, anthropometric data, smoking habit, constitutional susceptibility), psychosocial factors, and previous back traumas are also recognized as important predictors for low back pain (LBP), (Bongers and Boshuizen 1990). Therefore, injuries in the lower back of professional drivers represent a complex of health disorders of multifactorial origin involving both occupational and non-occupational factors. As a result, it is hard to separate the contribution of WBV exposure to the onset and the development of low back troubles from that of other individual and ergonomic risk factors. This makes it also difficult to establish exposure–response relationships for low back disorders among WBV exposed workers.

7.2.2 Low Back Disorders and WBV Exposure

The long-term effects of WBV on the lumbar spinal system have been reviewed by various authors (Bongers and Boshuizen 1990). In 1987, Hulshof and Veldhuijzen van Zanten scored the quality of exposure data, health effect data, and study design and methodology of 19 papers or reports which investigated locomotive thoracic and lumbar disorders in driving occupations. They concluded that the available epidemiologic data indicated a positive association between WBV exposure and spinal disorders but were inadequate to draw exposure-response relations.

More recently, Bovenzi and Hulshof (1998) have updated the information on the epidemiologic evidence of the adverse health effects of WBV on the spinal system by means of a review of the epidemiologic studies published between 1986 and 1997. In a systematic search, using several databases, of epidemiologic studies of LBP disorders and occupations with exposure to WBV, 45 articles were retrieved. The quality of each study was evaluated according to criteria concerning the assessment of vibration exposure, assessment of health effects, and methodology. A meta-analysis was also conducted in order to combine the results of independent epidemiologic studies. After applying the selection criteria, 17 articles reporting the occurrence of LBP disorders in 22 WBV exposed occupational groups, reached a sufficient score. The study design was cross-sectional for 15 occupational groups, longitudinal for 6 groups and of case-control type for one group. The main reasons for the exclusion of studies were insufficient quantitative information on WBV exposure and the lack of control groups.

Crane operators, bus drivers, tractor drivers, and fork-lift truck drivers were the most frequently investigated occupational groups in either cross-sectional or cohort studies. The control groups included in the epidemiologic studies consisted of either sedentary workers such as administrative officers or manual workers such as maintenance operators.

In the majority of the studies, vibration measurements on the vehicles were performed according to the international standard ISO 2631-1. The reported values of vibration magnitude (vector sum of the frequency weighted r.m.s. accelerations) varied from 0.25 to $0.67\,m\,s^{-2}$ in cranes, 0.36 to $0.56\,m\,s^{-2}$ in buses, 0.35 to $1.45\,m\,s^{-2}$ in tractors, and 0.79 to $1.04\,m\,s^{-2}$ in fork-lift trucks and freight-container tractors.

For the assessment of health effects in cross-sectional studies, the investigators used predominantly medical interview or questionnaires identical or similar to the standardized Nordic Questionnaire on musculoskeletal symptoms (Bongers and Boshuizen 1990). In most cases, the occurrence of LBP was investigated with respect to lifetime and the past 12 months. Medical records, providing information on the results of clinical and/or radiological investigations, were the basic data source of cohort studies of disorders of the spinal system, including lumbar intervertebral disc disorders.

The findings of both cross-sectional and cohort epidemiologic studies indicated that there is an increased risk for LBP disorders among occupational groups exposed to WBV when compared to unexposed control groups. Most of the reviewed studies reported risk estimates for LBP disorders that were adjusted for several confounders linked to individual characteristics (e.g. age, anthropometric variables, smoking, education) and other ergonomic risk factors (e.g. heavy physical work, lifting, twisting and bending). Psychological risk factors at work, such as perceived mental stress and job dissatisfaction, were also taken into account in several cross-sectional studies. Some trend of increasing prevalence of LBP disorders with the increase of WBV exposure was observed in cross-sectional studies of bus drivers, tractor drivers, fork-lift truck drivers and wheel loaders, as well as in a case-control study of disability pensioning due to degenerative changes in the spine of drivers of the transportation industry.

The results of the meta-analysis confirmed the findings of individual studies. A significant increase in the combined prevalence odds ratio (POR) for 12-month LBP was found in occupations with exposure to WBV from industrial vehicles (Figure 7.1).

An excess risk for sciatic pain and lumbar disc disorders, including herniated disc, was also found in the WBV exposed occupational groups compared with the control groups. It is worth noting that, with regard to herniated lumbar disc, the findings of the meta-analysis of cross-sectional studies (summary POR: 1.5; 95% CI: 0.9–2.4) were consistent with those of the meta-analysis of cohort studies (summary incidence density ratio: 1.8; 95% CI: 1.1–3.1).

Thus, the findings of the selected studies and the results of the meta-analysis of both cross-sectional and cohort studies confirmed that

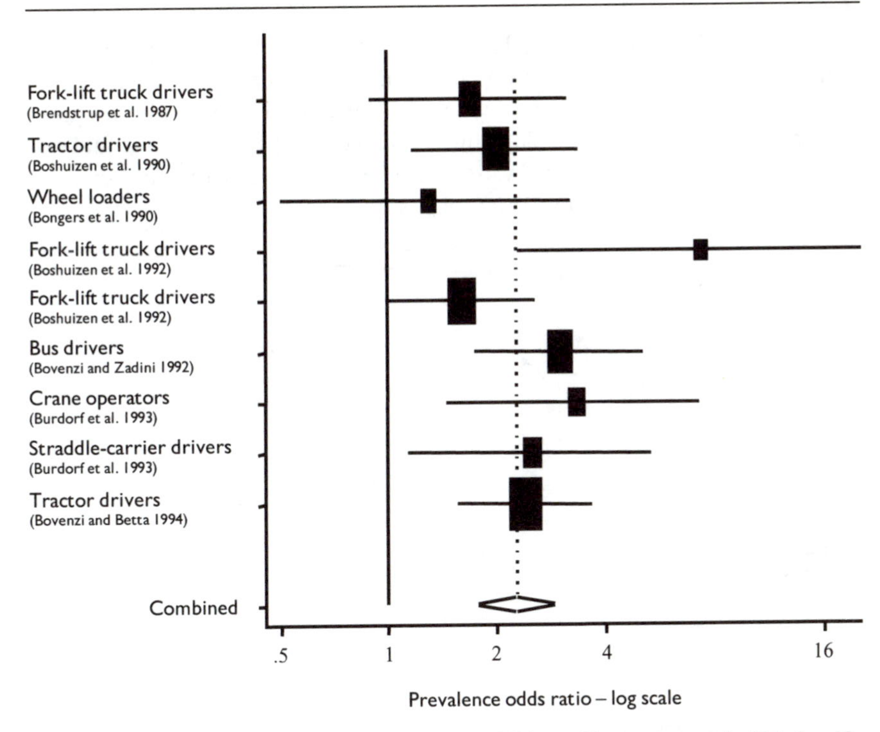

Figure 7.1 Prevalence odds ratios (POR) and 95% confidence intervals (CI) for 12-month low back pain in nine driving occupations with exposure to whole-body vibration compared to control groups. The area of each box is inversely proportional to the estimated effect's variance in the study. Random effects estimation of the combined POR and 95% CI is shown (see Bovenzi and Hulshof 1998 for details and references to the original studies).

occupational exposure to WBV is associated with an increased risk for LBP, sciatic pain, and degenerative changes in the spinal system, including lumbar intervertebral disc disorders. However, owing to the cross-sectional design of the majority of the reviewed studies, this epidemiologic evidence is not sufficient to outline a clear exposure-response relationship between WBV exposure and LBP disorders. Nevertheless, some elements of exposure–response relationship may be derived from two epidemiologic studies included in the review (Bongers and Boshuizen 1990; Bovenzi and Hulshof 1998). These studies, which investigated large samples of tractor drivers, are to a great extent comparable. The investigators used the same methods to measure WBV at the workplace and to assess cumulative vibration dose according to the equal energy principle. The two tractor driver groups differed with respect to mean duration of exposure (10 vs 21 years) and vibration magnitude (0.7 vs 1.1 m s^{-2}). LBP symptoms were collected with a similar questionnaire and the influence of potential confounders and postural load was taken into account in the study design or data analysis. Figure 7.2 displays the estimated POR for LBP as a function of the lifetime cumulative WBV dose, suggesting a trend for an increasing risk for LBP with increasing WBV exposure.

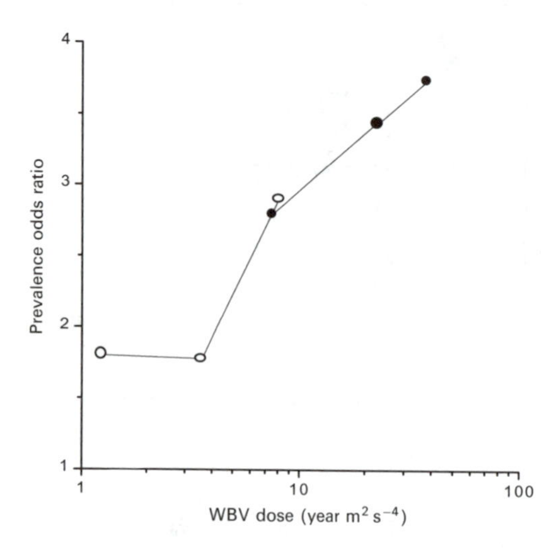

Figure 7.2 Prevalence odds ratio for low back pain among tractor drivers as a function of lifetime cumulative whole-body vibration (WBV) dose estimated as $\sum a_{vi}^2 t_i$, where a_{vi} is the vector sum of the frequency weighted root mean square acceleration of tractor i and t_i is the number of full-time working years driven on tractor i (year m^2 s^{-4}). ○ Study of Boshuizen et al. 1990, *International Archives of Occupational and Environmental Health*, 62, 109–115; ● Study of Bovenzi and Betta 1994, *Applied Ergonomics*, 25, 231–241.

In conclusion, there is clear evidence for an increased risk for LBP disorders in occupations with exposure to WBV. Biodynamic and physiological experiments have shown that seated WBV exposure can affect the spine by mechanical overloading and excessive muscular fatigue, supporting the epidemiologic findings of a possible causal role of WBV in the development of (low) back troubles. The fact that the WBV measured in most of the industrial vehicles involved in the review exceed the 8-hour action value of 0.5 m s^{-2}, and even the exposure limit value of 0.7 m s^{-2}, proposed by a recent European directive (Council of the European Union, 1994), stresses the relevance of the problem. This should stimulate the adoption of technical and health measures in order to prevent the onset of adverse health effects on the spine of drivers. Guidelines for the protection of workers against WBV exposure are included in the CEN report CR

12349:1996, and the European standards EN 30326-1 : 1994 and EN 1299 : 1997.

7.3 HAND-TRANSMITTED VIBRATION

7.3.1 General

Prolonged exposure to hand-transmitted vibration from powered processes or tools is associated with an increased occurrence of symptoms and signs of disorders in the vascular, neurological and osteoarticular systems of the upper limbs (Griffin 1990; Bovenzi 1997). The complex of these disorders is called *hand-arm vibration (HAV) syndrome*. The vascular component of the HAV syndrome is represented by a secondary form of Raynaud's phenomenon known as vibration induced white finger (VWF); the neurological component is characterized by a peripheral, diffusely distributed neuropathy with predominant sensory impairment; the osteoarticular component includes degenerative changes in the bones and joints of the upper extremities, mainly in the wrists and elbows. An increased risk for upper limb muscle and tendon disorders, as well as for nerve trunk entrapment syndromes, has also been reported in workers who use hand held vibrating tools (NIOSH, 1997). The vascular and osteoarticular disorders caused by hand-transmitted vibration are included in a European schedule of recognized occupational diseases (Commission of the European Communities 1990). It is estimated that 1.7 to 3.6% of the workers in the European Countries and the U.S. are exposed to potentially harmful hand-transmitted vibration (CR 12349 : 1996).

7.3.2 Neurological disorders

There is epidemiologic evidence for a greater occurrence of digital paraesthesias and numbness, deterioration of finger tactile perception, and loss of manipulative dexterity in occupational groups using vibrating tools than in control groups not exposed to hand-transmitted vibration (Griffin 1990). Epidemiologic surveys of vibration exposed workers have shown that the prevalence of peripheral sensorineural disorders varies from a few percent to more than 80%, and that symptoms and signs of sensory loss can affect users of a wide range of tool types (Bovenzi 1997).

Neurophysiological studies have suggested that sensory disturbances in the hands of vibration exposed workers are likely due to vibration induced impairment to various skin mechanoreceptors (Meissner's corpuscles, Pacinian corpuscles, Merkel cell neurite complexes, Ruffini endings) and their afferent nerve fibres. In acutely exposed animals, vibration can induce perineurial oedema, followed by fibrous thickening of the perineurium. Electron microscopic studies of human finger biopsy specimens suggest that hand-transmitted vibration can provoke perineural fibrosis, demyelination, axonal degeneration and nerve fibre loss.

The neurological component of the HAV syndrome is currently staged according to the scale proposed at the Stockholm Workshop 86 (1987). The sensorineural scale consists of three stages according to the symptoms complained and the results of clinical examination and objective tests (Table 7.1).

Table 7.1 Sensorineural stages of the hand-arm vibration syndrome according to the Stockholm scale.

Stage	*Symptoms and signs*
0SN	Exposed to vibration but no symptoms
1SN	Intermittent numbness, with or without tingling
2SN	Intermittent or persistent numbness, reduced sensory perception
3SN	Intermittent or persistent numbness, reduced tactile discrimination and/or manipulative dexterity

Clinical and epidemiologic surveys have revealed an increase in aesthesiometric, thermal, or vibrotactile perception thresholds of fingertips with the increase of daily vibration exposure, duration of exposure, or lifetime cumulative vibration dose. On the basis of these findings, various authors have discussed the possible form of an exposure–response relationship for vibration induced sensorineural disorders (Nilsson 1998). The results of some investigations suggested a tentative proposal of exposure–response relationship in which symptoms and signs of sensorineural disorders are likely to appear earlier than vascular disorders, even though these latter seem to develop more rapidly after their onset. Moreover, some longitudinal studies have shown that symptoms and signs of sensorineural abnormalities are more resistant to improvement or recovery after stopping the use of vibrating tools than vascular disorders (Bovenzi 1997; Nilsson 1998). These findings support the notion that there are different pathogenic mechanisms for nerve and vascular injuries caused by hand-transmitted vibration and that these disorders tend to develop independently of each other and at different rates. However, the currently available epidemiologic data are insufficient to outline the form of a possible exposure–response relationship for vibration induced neuropathy owing to the unspecific character of the sensory disturbances, uncertainties about the clinical validity of the Stockholm scale, the cross-sectional design of most epidemiologic studies, as well as the confounding and/or modifying effects of some variables linked to individual characteristics (age, alcohol consumption, body constitution) and diseases affecting the peripheral nervous system (metabolic disorders, injuries of the cervical spine, polyneuropathies).

Some cross-sectional and case-control studies have shown an increased occurrence of symptoms and signs of entrapment neuropathies, mainly carpal tunnel syndrome (CTS), in occupations involving the usage of vibrating tools (NIOSH 1997; Nilsson 1998). CTS is also common in job categories whose work tasks involve high force and repetitive hand–wrist movements. The independent contribution of vibration exposure and physical work load (forceful gripping, heavy manual labour, wrist flexion and extension), as well as their interaction, in the etiopathogenesis of CTS have not yet been established in epidemiologic studies of workers handling vibratory tools. It has been suggested that ergonomic risk factors are likely to play the dominant role in the development of CTS. As a result, to date it is hard to draw a specific relation between CTS and exposure to hand-transmitted vibration.

7.3.3 Bone and joint disorders

Vibration induced bone and joint disorders are a controversial matter. Various authors consider that disorders of bones and joints in the upper extremities of workers using hand-held vibrating tools are not specific in character and similar to those due to the ageing process and to heavy manual work (Gemne and Saraste 1987). Early radiological investigations had revealed a high prevalence of bone vacuoles and cysts in the hands and wrists of vibration exposed workers, but more recent studies have shown no significant increase with respect to control groups made up of manual workers (Gemne and Saraste 1987; Bovenzi 1997). An increased risk for wrist osteoarthrosis and elbow arthrosis and osteophytosis has been reported in coal miners, road construction workers and metal-working operators exposed to shocks and low frequency vibration (<50 Hz) of high magnitude from percussive tools (pick, riveting and chisel hammers, vibrating compressors), (CR 12349 : 1996). An excess prevalence of Kienbock's disease (lunate malacia) and pseudoarthrosis of the scaphoid bone in the wrist has also been reported by a few investigators (Gemne and Saraste 1987). On the contrary, there is little evidence for an increased prevalence of degenerative bone and joint disorders in the upper limbs of workers exposed to mid- or high-frequency vibration arising from chainsaws or grinding machines. It is thought that, in addition to vibration, joint overload due to heavy physical effort, awkward postures and other biomechanical factors can account for the higher occurrence of skeletal injuries found in the upper limbs of users of percussive tools (Gemne and Saraste 1987; Bovenzi 1998). A constitutional susceptibility might also play a role in the etiopathogenesis of premature wrist and elbow osteoarthrosis. At present, there are no epidemiologic studies that may suggest, even tentatively, an exposure–response relation for bone and joint disorders in vibration exposed workers.

7.3.4 Muscle and tendon disorders

Workers with prolonged exposure to vibration may complain of muscular weakness, pain in the hands and arms, and diminished muscle force (Griffin 1990; CR 12349 : 1996). Vibration exposure has also been found to be associated with a reduction of handgrip strength. In some individuals muscle fatigue can cause disability. Direct mechanical injury or peripheral nerve damage have been suggested as possible aetiologic factors for muscle symptoms. Other work related disorders have been reported in vibration exposed workers, such as tendinitis and tenosynovitis in the upper limbs, and Dupuytren's contracture, a disease of the fascial tissues of the palm of the hand (CR 12349 : 1996; NIOSH 1997). These disorders seem to be related to ergonomic stress factors arising from heavy manual work, and the association with handtransmitted vibration is not conclusive.

7.3.5 Vascular disorders (white fingers)

VWF is recognized as an occupational disease in many industrialized countries. Epidemiologic studies have pointed out that the prevalence of VWF is very wide, from 0–5% in workers using vibratory tools in geographical areas with a warm climate to 80–100% in workers exposed to high vibration magnitudes in northern countries (CR 12349 : 1996).

It is believed that vibration can disturb the digital circulation making it more sensitive to the vasoconstrictive action of cold. To explain cold induced Raynaud's phenomenon in vibration exposed workers, some investigators invoke an exaggerated central vasoconstrictor reflex caused by prolonged exposure to harmful vibration, while others tend to emphasize the role of vibration induced local changes in the digital vessels (e.g. thickening of the muscular wall, endothelial damage, functional receptor changes). It has also been suggested that vasoactive substances, immunologic factors or blood viscosity may play a role in the pathogenesis of VWF.

Clinically, VWF is characterized by episodes of white fingers caused by spastic closure of the digital arteries. A blue discoloration of the fingers (cyanosis) may follow. The attacks are usually triggered by cold and last from 5 to 30–40 minutes. A complete loss of tactile sensitivity may be experienced during an attack. In the recovery phase, commonly accelerated by warmth or local massage, redness (hyperaemia) may appear in the affected fingers as a result of a reactive increase of blood flow in the cutaneous vessels. In the rare advanced cases, repeated and severe digital vasospastic attacks can lead to trophic changes (ulceration or gangrene) in the skin of the fingertips. A grading scale for the classification of VWF has been proposed at the Stockholm Workshop 86 (1987), consisting of four symptomatic stages, from mild (1) to very severe (4). VWF symptoms are staged according to the frequency of finger blanching attacks, the number of

Table 7.2 The Stockholm Workshop scale for staging cold induced Raynaud's phenomenon in the hand-arm vibration syndrome.

Stage	*Grade*	*Symptoms*
0	–	No finger blanching attacks
1	Mild	Occasional attacks affecting only the tips of one or more fingers
2	Moderate	Occasional attacks affecting distal and middle (rarely also proximal) phalanges of one or more fingers
3	Severe	Frequent attacks affecting all phalanges of most fingers
4	Very severe	As in stage 3, with trophic skin changes in the fingers

affected fingers and the number of affected phalanges in a given finger (Table 7.2). The staging is made separately for each hand. A method of scoring the areas of the fingers affected with VWF has been developed to overcome some of the problems of staging VWF symptoms (Griffin 1990).

According to the resolutions of the Stockholm Workshop 94 (1995), a medical interview is generally accepted to be the best available method of reference to diagnose VWF. The following minimal requisites for the diagnosis of currently active VWF in a medical interview have been suggested: (i) cold provoked episodes of well demarcated acral blanching/whiteness in one or more fingers (a history of only cyanosis is not accepted as VWF); (ii) first appearance of finger blanching after start of professional exposure to hand-transmitted vibration and no other probable causes of Raynaud's phenomenon; (iii) VWF is currently active if episodes of blanching have been noticed during the last two years. If no episodes have occurred the VWF has ceased, provided there has been no change in cold exposure. Careful clinical investigation is needed, mainly for medico-legal purposes, to exclude the possibility that vibration exposed patients are affected with primary Raynaud's phenomenon (constitutional white finger) or with other secondary forms of Raynaud's phenomenon such as those caused by connective-tissue diseases, local trauma, compression of vessels (e.g. thoracic outlet syndrome), occlusive vascular diseases, neurological or immunological disorders.

Several laboratory tests are used to diagnose VWF objectively. Most of these tests are based on cold provocation and the measurement of finger skin temperature or digital blood flow and pressure before and after cooling of fingers and hands. Visual inspection of the change in the skin colour of the fingers after local cooling is the simplest method used by investigators to detect a blanching attack. Thermometric and plethysmographic methods are also frequently used methods for assessing the digital vasoconstrictor response to cold stimulation. It should be noted that a positive cooling test supports the diagnosis of VWF made during interview, but a negative test does not exclude the diagnosis.

The association between VWF and occupations involving work with vibrating tools has been clearly established in epidemiologic studies (Griffin 1990; CR 12349 : 1996; Bovenzi 1997). Several investigators have reported data that indicate a trend for an increasing occurrence of VWF with the increase of the magnitude of hand-transmitted vibration, the duration of exposure, or various measures of cumulative vibration dose obtained from combining vibration magnitude (frequency weighted or unweighted accelerations) and exposure time (years of tool usage or total working hours) (Bovenzi 1998).

Since the late 1970s a decrease in the occurrence of VWF has been reported among active forestry workers in both Europe and Japan after the introduction of antivibration chainsaws and administrative measures curtailing the saw usage time together with endeavours to reduce exposure to other harmful work environment factors (e.g. cold and physical stress). Recovery from VWF has also been reported among retired forestry workers.

An overview of the available epidemiologic literature seems to indicate a tendency towards a decrease in the prevalence of VWF in the last decade, at least among occupational groups who started to work with vibrating tools of new generation. Even though the magnitude of the reduction of VWF risk cannot be precisely estimated, there are reasons to believe that the exposure–response relationship proposed by the international standard ISO 5349 in 1986 needs to be updated as it was derived from investigations carried out from 1946 to 1978. It may be supposed that the findings reported in those studies are not representative of the exposure and health conditions associated with current work with vibrating tools.

A revision of the standard ISO 5349 is currently in preparation in an ad hoc ISO subcommittee (ISO/TC 108/SC 4/WG 3). In the revised ISO standard, the exposure–response relationship is restricted to a 10% prevalence of VWF and the probability of developing finger blanching symptoms in such a percentage of exposed workers is modelled as a function of the 8-hour energy equivalent frequency weighted acceleration sum, $a_{hv(\mathrm{eq,8h})}$ [or A(8)] in m s^{-2}, and the group mean total (lifetime) exposure duration, D_y in years.

The new proposal of ISO exposure–response relationship is very similar to that derived from the results of a recent epidemiologic study of forestry workers (Bovenzi 1998) in which the following power relation could be estimated (Figure 7.3):

$$P(\mathrm{VWF}) = 0.354 \times (\mathrm{A}(8))^{1.05} \times (D_y)^{1.07} \quad \% \qquad (7.1)$$

The fitted model indicates that the expected percentage of workers affected with VWF (P(VWF)) will vary roughly linearly with either A(8) (with D_y unchanged) or total exposure duration (with equivalent acceleration unchanged).

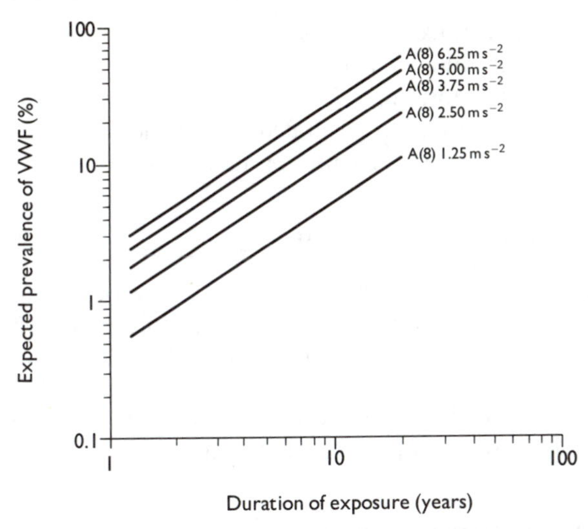

Figure 7.3 Expected prevalence of vibration induced white finger (VWF) in a population of 222 forestry workers as a function of 8-hour energy equivalent frequency weighted acceleration (A(8)) and the total duration of vibration exposure (Bovenzi 1998).

The findings of this study, therefore, suggest a relatively simple exposure-response relationship such that, if vibration magnitude is doubled, a halving of the years of exposure is required to produce the same effect. It should be noted that this prediction model is restricted to specific exposure conditions (chainsaw work) and the extrapolation to different occupational groups may not reflect the actual risk of adverse health effects arising from other types of vibration exposure. For instance, the findings of an epidemiologic study of operators using percussive stone working tools (Bovenzi 1998) suggested an exposure-response relationship in which the expected occurrence of VWF tended to increase roughly in proportion to the square root of A(8) (for a particular exposure period) or in proportion to the square root of the duration of exposure (for a vibration of constant magnitude). Thus, the two studies suggest the same time-dependency for the risk of VWF, but the magnitude of the effect for the same vibration exposure (intensity and duration) is different when the predicted occurrence of VWF in the stone workers is compared to that estimated in the forestry workers. This may be due to differences in the exposure conditions between the two occupational groups. Such differences might be connected with the physical characteristics of vibration (tools with dominant low- or high-frequency vibration, tools producing shock-type vibration), the work method (extent of the push and grip forces exerted on tool handles), and the pattern of daily exposure (continuous or intermittent work, short or long break periods). Moreover, the current ISO 5349 frequency weighting curve implies that the adverse health effects of hand-transmitted vibration decrease with the increase of the frequency of vibration, but this assumption may be inappropriate for certain types of vibration. The application of different frequency weighting methods to vibration generated by specific tool types should be considered for the assessment of vibration-induced disorders.

7.3.6 Other possible vibration-induced disorders

A few clinical and epidemiologic studies have reported that exposure to hand-transmitted vibration can aggravate the risk of noise induced hearing loss and provoke disturbances of the central nervous system (Griffin 1990; CR 12349 : 1996). To date, no exposure–response relationship can be derived from the findings of the studies which have investigated these disorders in occupational groups operating vibratory tools.

7.3.7 Exposure values and protection from hand-transmitted vibration

Although a reliable exposure–response relationship is an important requirement for the definition of vibration exposure limits, nevertheless the current inaccuracies in the relation between hand-transmitted vibration and its effects should not impede to provide some guidances to restrict hazardous vibration exposure at the workplace in order to guarantee the health and safety protection of workers. The Council of the European Union (1994) has suggested exposure values for hand-transmitted vibration expressed in terms of A(8). In the EU proposal, the action value is established at $2.5\,m\,s^{-2}$ and the exposure limit value at $5\,m\,s^{-2}$. The suitability of these exposure values was evaluated on the basis of the results of the previously mentioned study of forestry workers (Bovenzi 1998). Table 7.3 reports the number of years of vibration exposure that are expected to cause VWF in selected percentiles of

Table 7.3 Number of years of exposure for the onset of vibration-induced white finger (VWF) in various percentages of forestry workers according to the vibration exposure values proposed by the Directive of the Commission of the European Union (EUC). Daily exposure to hand-transmitted vibration is expressed in terms of 8-hour energy equivalent frequency weighted acceleration [A(8)]. The predicted years of exposure before VWF are estimated by means of equation (1) (Bovenzi 1998).

		Percentage of forestry workers affected with VWF				
EUC value	*A(8) ($m\,s^{-2}$)*	*10*	*20*	*30*	*40*	*50*
Threshold	1	22.5	42.9	>45.0	>45.0	>45.0
Action	2.5	9.2	17.5	25.5	33.3	41.1
Exposure limit	5	4.6	8.9	12.9	16.9	20.8

a worker population according to the EU exposure values. The predicted years of vibration exposure before the onset of VWF were estimated by equation (1). The estimated magnitude of the risk of VWF among workers exposed to A(8) of 2.5 or $5\,m\,s^{-2}$ indicates that the exposure values proposed by the EU directive are not too conservative.

The prevention of injuries or disorders caused by handtransmitted vibration requires the implementation of administrative, technical and medical procedures. Guidelines on preventive procedures are included in the European standard ENV-ISO 25349:1992 and in the CEN Reports CR 1030 : 1995 (Parts 1 and 2) and CR 12349:1996. Prevention includes technical measures aimed at elimination or reduction of hand-transmitted vibration at the source, appropriate information and advice to employers and employees, instruction to adopt safe and correct work practices, and medical preventive guidance. The presently available personal protective equipment against hand-transmitted vibration is of limited adequacy. Gloves are useful to protect the fingers and hands from traumas and to keep them warm. To be effective at attenuating vibration, gloves must pass the test required by the international standard ISO 10819 : 1996.

7.4 ACKNOWLEDGEMENT

This study was supported by the European Commission under the BIO-MED 2 concerted action BMH4-CT98-3251 (Vibration Injury Network).

7.5 REFERENCES

Bongers P, Boshuizen H. *Back disorders and whole-body vibration at work*, Ph.D. thesis, University of Amsterdam, 1990.

Bovenzi M. Hand-transmitted vibration. In *Encyclopaedia of Occupational Health & Safety*, Volume 2, Part IV, Chapter 50: Vibration, 4th ed., edited by JM Stellman. International Labour Office, Geneva, 1997: 50.7–50.12.

Bovenzi M, Hulshof CTJ. An updated review of epidemiologic studies on the relationship between exposure to whole-body vibration and low back pain. *J Sound Vibration* 1998; 215, 595–611.

Bovenzi M. Exposure-response relationship in the hand-arm vibration syndrome: an overview of current epidemiology research. *Int Arch Occupational Environmental Health* 1998; 71, 509–519.

CEN. *Mechanical vibration – Guide to the health effects of vibration on the human body*. Comité Européen de Normalisation, Brussels. Reference no. CR 12349 : 1996.

Commission of the European Communities. Commission recommendation of 22 may 1990 to the Member States concerning the adoption of a European schedule of occupational diseases. *Official J European Communities* 1990; 90/326/EEC, no. L 160/39(48).

Council of the European Union. Amended proposal for a Council Directive on the minimum health and safety requirements regarding the exposure of workers to the risks arising from physical agents – Individual Directive in relation to Article 16 of the Directive 89/391/EEC. *Official J European Communities* 1994; 94/C 230/03, no. C 230/3(29).

Gemne G, Saraste H. Bone and joint pathology in workers using hand-held vibrating tools. An overview. *Scand J Work Environ Health* 1987; B13, 290–300.

Griffin MJ. *Handbook of Human Vibration*. Academic Press, London, 1990.

Hulshof CTJ, Veldhuijzen van Zanten OBA. Wholebody vibration and low back pain – a review of epidemiologic studies. *Int Arch Occupational Environmental Health* 1987; 59, 205–220.

ISO. *Mechanical vibration – Guidelines for the measurement and the assessment of human exposure to handtransmitted vibration*. International Organization for Standardization, Geneva. Reference no. ISO 5349 : 1986.

ISO. *Guide for the evaluation of human exposure to whole body vibration – Part 1: General requirements*. International Organization for Standardization, Geneva. Reference no. ISO 2631-1 : 1997.

National Institute for Occupational Safety and Health. *Musculoskeletal Disorders and Workplace Factors*, edited by BP Bernard. DHHS, NIOSH, Washington, DC. Publication no. 97(141), 1997.

Nilsson T. Neurosensory function and white finger symptoms in relation to work and hand-transmitted vibration. *Arbete och Hälsa* 1998; 29, 1–67.

Stockholm Workshop 86. Symptomatology and diagnostic methods in the hand-arm vibration syndrome. *Scand J Work Environ Health* 1987; 13 (no. 4, special issue), 265–388.

Stockholm Workshop 94. Hand-arm vibration syndrome: diagnostics and quantitative relationships to exposure. *Arbete och Hälsa*, 1995, 5, 1–199.

Chapter 8

Screening for Upper Extremity Musculoskeletal Disorders: Research and Practice

Alfred Franzblau

8.1 INTRODUCTION

> Various and manifold is the harvest of disease reaped by certain workers from the crafts and trades that they pursue; all the profit that they get is fatal injury to their health...mostly from two causes. The first and most potent is the harmful character of the materials that they handle ... the second cause I ascribe to certain violent and irregular motions and unnatural postures of the body, by reason of which the natural structure of the vital machine is so impaired that serious diseases gradually develop therefrom.
>
> Ramazzini (1713)

As suggested by the 'second cause' identified by Dr Ramazzini in the year 1713, musculoskeletal disorders related to employment have been a major problem for centuries. However, it has only been in the last 15 or 20 years that upper extremity musculoskeletal disorders (UEMSDs), and carpal tunnel syndrome (CTS) in particular, have gained wide recognition as major medical problems among industrial and office workers in the United States and other industrialized countries (Hanrahan LP 1991; Bureau of Labor Statistics, 1995). In the mid-1980s the United States National Institute for Occupational Safety and Health (NIOSH) identified workrelated musculoskeletal disorders as among the ten most important occupational safety and health concerns in the United States (NIOSH 1986).

8.2 BACKGROUND

As with other public health problems in general, and work-related diseases in particular, disease surveillance is one important tool for understanding and ultimately controlling the problem. A surveillance programme should include case counting, data evaluation, and a plan of action (Baker 1989; Landrigan 1989).

'Screening' usually implies a programme of medical history, physical examination, and/or laboratory tests designed to detect a specific disease process at an early, potentially reversible stage. A screening programme is one form of case counting for the purposes of surveillance. Testing the blood of lead-exposed workers for lead is an example of a widely accepted medical screening programme.

There are many potential reasons for conducting medical screening of workers for UEMSDs. Examples of objectives for conducting medical screening among workers may include:

- To identify causal determinants of particular work-related diseases;
- To identify jobs with high rates of disorders so that interventions can be targeted;
- To determine if a control programme has been effective in reducing incidence, severity or cost of disease;
- To identify individual workers with possible disease so they may obtain appropriate referral and early therapy;
- To predict future risk of development of disease in individual workers so as to positively influence appropriate job placement;
- To assess job compatibility of individual workers.

It may seem obvious, but before embarking on a screening programme, the goals and objectives of the programme must be stated clearly and understood. And, although 'success' can be defined in numerous ways, it only makes sense that the screening programme should have a reasonable chance of achieving the prescribed goals and objectives, however defined.

Surveillance techniques generally can be classified as either 'passive' or 'active'. 'Passive' surveillance involves utilization of data bases which were designed or implemented for some reason other than to directly monitor occupational disease or injury in the workplace. For example, Workers' Compensation records are maintained primarily to monitor employer costs, but review of such records may yield valuable information pertaining to patterns or severity of injury and/or illness in a particular workplace. Usually, passive surveillance techniques are less expensive because no additional expenditure is required to collect the data. However, they also suffer from a number of deficiencies, the most significant being underreporting and a lack of standardized definitions of what are UEMSDs.

Examples of tools used to perform passive surveillance for UEMSDs in the workplace include:

- OSHA 200 logs or other government-required records of workplace injuries
- Workers' Compensation Claims
- Medical Absences

- Plant Medical Records
- Workers' Complaints
- Workers' Personal Physician Diagnoses

'Active' surveillance or screening implies design and implementation of a programme specifically for the purpose of monitoring the incidence or prevalence of UEMSDs in the workplace. Such programmes may be more expensive to implement when compared to passive surveillance programmes. However, they have the advantage of more complete reporting, and also utilizing standardized and/or consistent case definitions of UEMSDs.

In order to be able to justify expenditure of resources related to implementing a medical screening programme for UEMSDs, it will usually be necessary to assess the likelihood of success. In particular, if the problem is common and/or prevalent, it is expensive, the proposed screening test(s) are cheap/reliable/valid, and interventions are possible, then circumstances will probably favour implementation of a medical screening programme.

There is no question that work-related UEMSDs (or their symptoms) are common, and the related costs are considerable (Hanrahan 1991; Bureau of Labor Statistics 1995; American Academy of Neurology 1993; Stevens 1988; Levine 1993; Franzblau 1993). The present talk will therefore focus on methodological and conceptual issues related to validity when screening for UEMSDs in general, and carpal tunnel syndrome (CTS) in particular.

8.3 EVALUATION OF SCREENING TOOLS

The first decision one must face when designing and implementing an active screening programme for CTS is what tests to choose. The most obvious choices are to employ adaptations of procedures commonly used in clinical practice for diagnosis of CTS: medical history/questionnaire surveys, physical examination procedures, and nerve conduction testing.

The medical history, particularly self-administered hand diagrams, have been identified as the best tool currently available for the purpose of eliciting symptom information pertaining to CTS (Rempel et al. 1998). Self-administered surveys are relatively cheap to administer and to interpret. Hand diagrams for screening for CTS have been described by Katz and Stirrat (1990) and by Katz et al. (1990), and later modified by Franzblau et al. (1994). Under actual field conditions (in which the survey results of individuals were confidential and not revealed to the employer), hand diagrams have been demonstrated to have good test-retest reliability (Franzblau et al. 1997).

However, a major criticism of hand diagrams, or any symptom survey, is the subjective nature of the information. Under certain circumstances, there is likely to be considerable underreporting of symptoms (i.e. in the pre-placement setting) (Bingham 1996), and one could imagine circumstances which might foster overreporting of symptoms. Hence, the utility of questionnairebased information, and hand diagrams in particular, for conducting a workplace survey is partly a function of the goal(s) of the survey, and also the circumstances under which the survey is administered (e.g., pre-placement examination versus an anonymous survey).

In the pre-placement setting, in which the employer's medical representative elicits information non-anonymously with the goal of impacting employment decisions, the likelihood of under (or over) reporting of symptoms may be enhanced. In contrast, if the survey is conducted anonymously (so the employee is assured of confidentiality of his/her responses), then the data are likely to be more reliable and valid. Obviously, with an anonymous survey, some goals may not be achievable (e.g., to identify individual workers with possible disease so they may obtain appropriate referral and early therapy), and so there is a practical trade-off.

The next category of possible screening tools to consider is physical examination procedures. Physical examination procedures for CTS usually require a trained examiner (and thus may be more expensive), but they are used widely and are relatively easy to administer. Examples includes Phalen's manoeuvre, Tinel's sign, the carpal compression test, and two-point discrimination. Despite the fact that such tests are administered by physicians and other health professionals, all such provocative manoeuvres have a subjective component which is often overlooked. An even more important issue relates to the validity and reliability of such procedures: how much useful information do these tests add in identifying CTS? Numerous studies have examined these issues, and results are largely concordant: despite the diversity of investigators, types of study designs, and study populations, there is remarkable agreement that these procedures are poorly suited for aiding in the clinical diagnosis (or screening) of CTS (Franzblau et al. 1993; Golding et al. 1986; DeKrom et al. 1990; Katz et al. 1990; Buch-Jaeger et al. 1994; Gerr and Letz 1998). Also, it should be noted that none of these provocative manoeuvres for CTS have an established relationship with job or task performance.

Finally, what about using nerve conduction studies (NCS) to screen for CTS? These studies require a skilled examiner and expensive equipment, but they are objective, widely used in clinical medicine, and can be highly reliable even under field conditions (Salerno et al. 1999). Hence, there are many attractive aspects to using NCS to screen for CTS. However, despite these potential advantages of using NCS in screening for CTS, there are major, usually unappreciated, limitations related to scientific validity and predictive value.

Recent studies have shown that NCS have only a weak relationship with typical clinical features of CTS, such as symptoms and physical examination findings (Homan 1999; Ferry 1998). Most persons with symptoms consistent with CTS do not have electrophysiologic abnormalities, and vice versa. This situation is analogous to the weak relationship between low back pain and abnormalities identified on magnetic resonance imaging of the low back Jensen et al. 1994).

The strong associations (but weak concordances) are probably acceptable for epidemiological purposes, but present great challenges in the clinical setting and when screening. In particular, certain screening goals may not be achievable (e.g. to identify individual workers with possible disease so they may obtain appropriate referral and early therapy.)

The results from such studies raise other questions: what about the large pool of workers who are asymptomatic, or do not describe symptoms typical of CTS, but who have abnormal NCS (between 15% and 25%, depending on the study population)? Are these people at substantially greater risk of going on to develop CTS? In a prospective study designed to address this question, we were surprised to find that this was not the case: asymptomatic workers with abnormal NCS were no more likely to develop symptoms of CTS during follow-up than asymptomatic workers with normal NCS matched on age, gender and place of employment (Werner et al. 1997). Hence, nerve conduction studies were not predictive of future development of symptoms consistent with CTS among asymptomatic workers. And, most workers with abnormal NCS did not report development of new-onset symptoms at follow-up. A key conclusion was that NCS are, at best, a poor tool for predicting future risk of CTS in individuals.

Interestingly, what about people who have symptoms of CTS, but who have normal NCS? This constellation of findings also represents a large fraction of the working population. Unfortunately, there are no prospective studies of this group to assess their natural history. However, there are at least two surgical case series which suggest that some of these people may benefit from carpal tunnel release surgery (Grundberg 1983; Hamanaka et al. 1995).

8.4 CONCLUSIONS AND RECOMMENDATIONS

In conclusion, it would appear that some of the underlying assumptions concerning screening for CTS (and other UEMSDs) have little empirical foundation. At present, there is no medical test or examination procedure which can predict future risk of CTS with a high enough degree of certainty to permit use in job placement decisions. Furthermore, NCS for CTS do not provide information about job compatibility or ability to perform particular job tasks. NCS allow for documentation of current clinical status or baseline with regard to nerve function. For other UEMSDs, the same is true: there are no medical tests which reliably can predict future risk of disease.

With these results in mind, the following recommendations on screening for UEMSDs among workers seem reasonable:

- Determine your goals/objectives for screening for UEMSDs – be realistic!
- Understand the circumstances under which you intend to collect your data.
- Assess whether your goals and objectives are feasible under the circumstances.
- Don't forget that the simplest tool is frequently the best: a short questionnaire is cheap and easy to design and administer, and can yield valid, reproducible information concerning populations of workers.
- Despite possible imperfections, don't ignore what may be 'free', such as OSHA 200 logs (or other government records), Workers' Compensation files, and plant medical records.

8.5 REFERENCES

American Academy of Neurology. Report of the Quality Standards Subcommittee of the American Academy of Neurology: Practice parameter for carpal tunnel syndrome (summary statement). *Neurology* 1993; 43: 2406–2409.

Baker EL, Honchar PA, Fine LJ. Surveillance in occupational illness and injury: Concepts and content. *Am J Pub Health* 1989; 79(suppl): 9–11.

Bingham RC, Rosecrance JC, Cook TM. Prevalence of abnormal median nerve conduction in applicants for industrial jobs. *Am J Industrial Medicine* 1996; 30(3): 355–361.

Buch-Jaeger N, Foucher G. Correlation of clinical signs with nerve conduction tests in the diagnosis of carpal tunnel syndrome. *J Hand Surg* 1994; 19B(6): 720–724.

Bureau of Labor Statistics. *Occupational Injuries and Illnesses in the United States by Industry*. US Department of Labor, Bureau of Labor Statistics, 1995.

DeKrom MCTFM, Knipschild PG, Kester ADM, Spaans F. Efficacy of provocative tests for diagnosis of carpal tunnel syndrome. *Lancet* 1990; 335: 393–395.

Ferry S, Silman AJ, Pritchard T, Keenan J, Croft P. The association between different patterns of hand symptoms and objective evidence of median nerve compression. *Arthritis Rheumatism* 1998; 41(4): 720–724.

Franzblau A, Salerno DF, Armstrong TJ, Werner RA. Test–retest reliability of an upper extremity discomfort questionnaire in an industrial population. *Scand J Work Environ Health* 1997; 23: 299–307.

Franzblau A, Werner R, Valle J, Johnston E. Workplace surveillance for carpal tunnel syndrome: A comparison of methods. *J Occup Rehab* 1993; 3(1): 1–14.

Franzblau A, Werner RA, Albers JW, Grant CL, Olinski D, Johnston E. Workplace surveillance for carpal tunnel syndrome using hand diagrams. *J Occup Rehab* 1994; 4(4): 185–198.

Gerr F, Letz R. The sensitivity and specificity of tests for carpal tunnel syndrome vary with the comparison subjects. *J Hand Surgery (Brit)* 1998; 23(2): 151–155.

Golding DN, Rose DM, Selvarajah K. Clinical tests for carpal tunnel syndrome: An Evaluation. *Brit J Rheum* 1986; 25: 388–390.

Grundberg AB. Carpal tunnel syndrome in spite of normal electromyography. *J Hand Surg (Am)* 1983; 8(3): 348–349.

Hamanaka I, Okutsu I, Shimizu K, Takatori Y, Ninomiya S. Evaluation of carpal canal pressure in carpal tunnel syndrome. *J Hand Surg* 1995; 20A: 848–854.

Hanrahan LP, Higgins D, Anderson H, Haskins L, Tai S. Project SENSOR: Wisconsin surveillance of occupational carpal tunnel syndrome. *Wis Med J* 1991; 90(2): 80, 82–83.

Homan MM, Franzblau A, Werner RA, Albers JW, Armstrong TJ, Bromberg MB. Agreement between symptom surveys, physical examination findings and electro-diagnostic testing for carpal tunnel syndrome. *Scand J Work Environ Health* 1999; 25(2): 115–124.

Jensen MC, Brandt-Zawadzki MN, Obuchowski N, Modic MT, Malkasian D, Ross JS. Magnetic resonance imaging of the lumbar spine in people without back pain. *N Engl J Med* 1994; 331: 69–73.

Katz JN, Stirrat CR. A self-administered hand diagram for the diagnosis of carpal tunnel syndrome. *J Hand Surg* 1990; 15A: 360–363.

Katz JN, Stirrat CR, Larson MG, Fossel AH, Eaton HM, Liang MH. A self-administered hand symptom diagram for the diagnosis and epidemiologic study of carpal tunnel syndrome. *J Rheumatol* 1990; 17: 1495–1498.

Katz JN, Larson MG, Sabra A, Krarup C, Stirrat CR, Sethi R, Eaton HM, Fossel AH, Liang MH. The carpal tunnel syndrome: Diagnostic utility of the history and physical examination findings. *Ann Int Med* 1990; 112: 321–327.

Landrigan PJ. Improving the surveillance of occupational disease. *Am J Pub Health* 1989; 79: 1601–1602.

Levine DW, Simmons BP, Koris MJ, Daltroy LH, Hohl GG, Fossel AH, Katz JN. A self-administered questionnaire for the assessment of severity of symptoms and functional status in carpal tunnel syndrome. *J Bone Joint Surg* 1993; 75A(11): 1585–1592.

National Institute for Occupational Safety and Health. *Proposed national strategy for the prevention of musculoskeletal injuries: DHHS (NIOSH) Publication No. 89-129*. Washington, D.C., 1986.

Rempel D, Evanoff B, Amadio PC, de Krom M, Franklin G, Franzblau A, Gray R, Gerr F, Hagberg M, Hales T, Katz JN, Pransky G. Consensus criteria for the classification of carpal tunnel syndrome in epidemiologic studies. *Am J Public Health* 1998; 88(10): 1447–1451.

Salerno DF, Werner RA, Albers JW, Becker MP, Armstrong TJ, Franzblau A. Reliability of nerve conduction studies among active workers. *Muscle Nerve* 1999; 22: 1372–1379.

Stevens JC, Sun S, Beard CM, O'Fallon WM, Kurland WT. Carpal tunnel syndrome in Rochester, Minnesota, 1961 to 1980. *Mayo Clin Proc* 1988; 38: 134–138.

Werner RA, Franzblau A, Albers JW, Buchele H, Armstrong TJ. Use of screening nerve conduction studies for predicting future carpal tunnel syndrome. *Occup Environ Med* 1997; 54: 96–100.

Chapter 9

Surveillance for musculoskeletal problems

Mats Hagberg

9.1 INTRODUCTION

Active health surveillance of work-related musculoskeletal disorders is one important tool to enhance workers' health and productivity. Today there are instruments, e.g. questionnaires that occupational health care professionals, consultants, and management can use in surveillance of worked-related musculoskeletal disorders to detect work tasks or work processes that constitute a hazard both to health and productivity (Morse 1999).

A surveillance system can be an integral part of an ergonomics programme, a part of a health and safety programme or a stand-alone activity.

9.2 WHAT IS SURVEILLANCE?

Surveillance is collecting and using information for action including planning, implementing and evaluating medical and ergonomic interventions (Halperin 1993; Klaucke et al. 1988; Fine 1999). Screening is defined as identification of unrecognized disease or defect by the application of tests, examinations or other procedures that can be applied rapidly. Screening tests sort out apparently persons who probably have a disease from those who probably do not (Last 1995). In health surveillance the primary objective is not to sort out people with a disorder but to get an estimate of the health of a population or a group and to link the health to group characteristics, i.e. exposure.

Both surveillance and research are motivated by prevention, however they are not synonymous (Halperin et al. 1992). Surveillance is the systematic ongoing collection, analysis and dissemination of information for the purpose of prevention interventions (Halperin 1993; Klaucke et al. 1988; Halperin et al. 1992). Research is an activity that focuses for a limited time period on a search for etiology, association or an evaluation of intervention. Research projects are sustained only long enough to collect sufficient information to test specific hypotheses in contrast to surveillance, which is an ongoing activity as long as there is interest in fulfilling one or more goals of surveillance. Surveillance systems collect limited information about cases, while research usually collects detailed information. Surveillance and research in the past have focused on disease and injury, and are only recently turning toward disability and exposure or hazard.

9.3 HAZARD VS. HEALTH SURVEILLANCE

A hazard is a potential risk that can cause an adverse health outcome if not dealt with in a proper way. To prevent musculoskeletal disorders we can focus on the hazards or health outcomes to assess the working environment (Table 9.1). Hazard surveillance may include collecting information concerning the job designs (example: job demand of power grips), the technologies used (example: interactive computing, vibrating machines), the organizations (example: team work, incentive pay) and the environment (example: cold climate). In health surveillance we usually think of collecting information about the diseases and disorders. However earlier signs of an unhealthy environment may be expressions of disorders sometimes called intermediate variables, for example ache and pain (Hagberg et al. 1995). Acute effects of exposure and even more sensitive variables may be recording information about comfort or discomfort (Saldana et al. 1994). Performance may also be a way of determine early health effects. Furthermore performance information may motivate the employer to initiate ergonomic improvements.

Table 9.1 Surveillance can be divided into hazard and health surveillance.

Surveillance	
Hazard	*Health*
Job/task design	Disorder/disease
Technology	Expressions of disorder/disease
Organization	Comfort
Environment	(Performance)

9.4 PASSIVE AND ACTIVE SURVEILLANCE

Passive surveillance uses existing data systems. Hazard surveillance may be done using data on the number of machines (for example, number of grinders), number of organizations (for example, workplaces with incentive pay in a big company) and knowledge about the environment (for example, number of hot/cold workplaces) (Table 9.2). Passive health surveillance can involve the existing data on occupational injuries for the workplace, absentee records and transfer requests to other jobs (Silverstein et al. 1997).

Active hazard surveillance implies designing and implementing a system where hazards are actively researched and evaluated. This may involve walkthroughs with checklists where exposure is evaluated (Pan et al. 1999), the use of questionnaires where the workers answer questions about exposures and even further the use of, for example, job analysis of videos. Active health surveillance typically uses questionnaires for symptom surveys. Periodically repeated physical exams may be used for surveillance and laboratory tests for example measurement of vibration perception threshold levels in vibratory exposed workers (Lundstrom et al. 1999). Probably future active health surveillance will involve biological monitoring of stress hormones and metabolites.

Table 9.2 Hazard and health surveillance can be divided into passive and active surveillance.

Passive		*Active*	
Hazard	*Health*	*Hazard*	*Health*
No. of machines	Occupational injury reports	Questionnaires	Questionnaires
No. of organizations	Absentee records	Checklists	Physical exams
No. of hot/cold workplaces	Transfer requests	Job analyses	Biological monitoring

9.5 GOALS OF SURVEILLANCE

9.5.1 Detection of new problems

Surveillance may detect the association of a musculoskeletal disorder with a new specific work process or occupation. This generally happens through two types of surveillance data: either the identification of cases without definite information about the size of the population the cases are drawn from (Sentinel Health Event); or from a surveillance source of cases which includes some information both on the number of cases and the size of the population at risk. Surveillance strategies are designed to recognize: (1) patterns of health and disease in groups of people and (2) the most obvious risk factor patterns in the workplace that may contribute to health/disease patterns.

> Example: The computer mouse has given new risks for symptom combinations of pain in wrist and in the shoulder as a 'syndrome' probably caused by exposure of non-neutral position of the wrist and constrained postures of the shoulder neck (Hagberg 1996). Surveillance can be used for assessing this problem.

9.5.2 Determine the magnitude of musculoskeletal disorders

The magnitude can be viewed as the number of cases measured as incidence or prevalence in the surveillance (Warren et al. 2000). The magnitude can also be viewed as the severity. The severity can be measured as workdays lost or reduced work performance.

9.5.3 Track trends over time

Increases or decreases in the number of cases or the changes in the rate of disorders may be due to changing levels of exposure or changes in the reporting of disorders independent of their level of occurrence. The tracking of trends can be done either at the national or local level (Fine 1999). Surveillance evaluations may involve both the levels of occupational exposures (hazard) and health outcome.

9.5.4 Identify occupational groups, work sites to target control measures

Surveillance can be used to identify occupational groups: departments and work site where problems occur to implement control measures. By surveillance there is evidence of the need for the control measure.

9.5.5 Describe health and risk factors to initiate ergonomic change

A feedback of surveillance information is essential both to management and the worker. The feedback of the result of the surveillance may initiate a discussion of possible control actions and implement ergonomic changes.

Feedback

Feedback is a method for changing people's behaviour. It has its background in behavioural-scientific research, and has been developed and utilized in a variety of projects. Feedback has been used for changing organizational behaviour. In industry feedback is used to increase productivity or quality of products. In occupational safety, feedback may be used for the reduction of accidents in working life. For example, by feeding back information on accidents, employees can be motivated to work in a safer manner (Menckel et al. 1997). The idea is that this way of working will lead to a reduction in the number of accidents.

Feedback may operate as a reward and/or punishment stimuli in itself. Goal accomplishment is an important indicator that is used in feedback. Feedback may also influence future performance goals. Behavioural feedback program consists of four phases. Determination phase where the safe work practice is established. Training phase where the workers are trained in safe work practice. Observation phase where the behaviours of work practice is observed. Provision phase where the feedback on the use of safe work practice is done.

Feedback of health surveillance results can be given for different levels. At the national level feedback of information from surveys can affect health and labour market policies proposed by the government. At regional and community level feedback of health surveillance can be the base for directing health resources. At the local level feedback of health surveillance can be given to the management and/or the employees. This section will deal with feedback of health surveillance data at the local level. There are four major questions concerning feedback of health surveillance data at the local level. These questions are: What information should be reported? When should the information be reported? How should the information be presented? To whom should the information be given?

What information should be reported?

When planning health surveillance it is important consider what information should be included in the report. There is no use in collecting information on topics that will not be used in a feedback situation. You have to consider what information you want to make feedback on when you decide what information to collect. It is unethical to collect information that is not to be used in the feedback and reporting situation. Since surveillance is not a research activity, information that is not known to be related to health or performance should not be collected. Since surveillance is an activity to promote changes, information that affects both health and performance can be effective to present. Presenting comparisons with earlier surveillance in the company or organization or presenting comparisons to other companies or organizations may initiate discussions about health, safety and the production system.

When should the feedback and reporting be performed?

The delay between collecting information and reporting information should be kept as short as possible. In passive health surveillance there might be no expectation of receiving information. However, in active health surveillance, those who participate and a management requesting the information are waiting and have expectations of receiving a report. Thus, in active health surveillance, the data collection should be made in a very short time period, 1–2 weeks, and the delay between the collection and the feedback should be aimed at two weeks. This implies that there has to be a system for data collection and analysis and making a report of the active surveillance data.

If there is a long delay between the data collection and the feedback situation the effect of initiating changes could be severely impaired. In a situation where there are a couple of months between the data collection and the feedback situation the employees and management will regard the information as old and not accurate for their present working situation.

How should the surveillance data be presented?

Feedback of surveillance data could be made in different ways. In its simplest way this could be a telephone conversation with the manager. In more advance ways it could be both oral and written reports to the management and workers. In a comparison of giving feedback of active surveillance data to the supervisors versus giving the feedback information to both supervisors and employees, more ideas were brought up for changes and more changes were implemented when feedback of active surveillance was made to both supervisors and employees (Menckel et al. 1997).

Even if an oral report is made to management or employees there is a benefit of also having a short written report. The written report will be a document that can be referred to and will also be evidence of what has taken place. The report should be short, one or two pages, so it will be read by management. Extensive reports will be filed without being read. In the report it is possible to point out specific health outcomes that were not expected or were different from the normal population. Generally the suggestions of improvement of the working place in the report of active surveillance data should be avoided. It is better to point out that it is the health – or performance – problems in the report and the offer as a conclusion to be readily available for consulting regarding measures and changes at the workplace. This way, especially if the active health surveillance is performed by occupational health centre there is a possibility to gain extra contracts as consultant for workplace changes. This could further develop the professional roles of the occupational health professionals.

To whom should the feedback and information be given?

As pointed out earlier, it is important to give the feedback information to both management and employees. If a choice has to be made between giving the information to the management or the employees, preference would be to give the information to the management. It is important to give the information in such a way that the management and employees understand the information.

9.5.6 Basis for prioritizing preventive actions

When the magnitude of musculoskeletal disorders are known for different occupational groups or work sites a priority for control actions can be made on the basis of number of cases or severity (Ricci et al., 1998).

9.5.7 Identification of control measures by observing low-risk groups

Observing and comparing low risk can identify groups with high-risk population potential control measures. The characteristics of low-risk population can be used as standard that may be transferred to the high-risk groups.

9.5.8 Evaluation of the progress of preventive actions

Tracking trends in the number of musculoskeletal disorders cases can be effective in evaluating the successfulness of an intervention programme to reduce the number of cases.

9.5.9 Generate hypothesis for research

Surveillance can generate hypothesis for research by revealing associations between exposure and health outcomes. Furthermore new syndromes or a combination of symptoms can trigger detailed research investigations.

There is a report by Klaucke (Fine 1999) on how to evaluate a surveillance system. The report can serve as a basis for establishing a quality assurance system for surveillance. With the rapid development of technology new methods in musculoskeletal health and hazard surveillance will develop. New positioning systems together with web cameras will provide opportunity to gather information for surveillance purposes.

9.6 REFERENCES

Fine LJ. Surveillance and occupational health. *Int J Occup Environ Health* 1999; 5(1): 26–29.

Hagberg M. Exposure considerations when evaluating musculoskeletal diagnoses, in *Advances in Occupational Ergonomics and Safety*, A. Mital et al. (eds). International Society for Occupational Ergonomics and Safety: Cincinnati, 1996: 411–415.

Hagberg M, Silverstein B, Wells R, Smith MJ, Hendricks HW, Caryon P, Pérusse M. In *Work Related Musculoskeletal Disorders (WMSDs): a reference book for prevention*, I Kuorinka, L Forcier, eds. London: Taylor & Francis Ltd, 1995: 1–421.

Halperin W. Occupational health surveillance. *Health and Environmental Digest* 1993; 8: 3–5.

Halperin W, Baker E, Monson R. *Public Health Surveillance*. Van Nostrand Reinhold, 1992.

Klaucke DN, Buehler JW, Thacker SB, Parrish RG, Trowbridge FL, Berkelman RL. Guidelines for evaluating surveillance systems. *MMWR* 1988; 37 (suppl.5): 1–18.

Last JM, ed. *A Dictionary of Epidemiology*, third edition. Oxford University Press: New York, 1995: 180.

Lundstrom R, Nilsson T, Burstrom L, Hagberg M. Exposure-response relationship between hand-arm vibration and vibrotactile perception sensitivity. *Am J Ind Med* 1999; 35(5): 456–464.

Menckel E, Hagberg M, Engkvist IL, Wigaeus Hjelm E. The prevention of back injuries in Swedish health care – a comparison between two models for action oriented feedback. *Applied Ergonomics* 1997; 28: 1–7.

Morse T. Surveillance and the problems of assessing office-related injury. *Occup Med* 1999; 14(1): 73–80, iii.

Pan CS, Gardner LI, Landsittel DP, Hendricks SA, Chiou SS, Punnett L. Ergonomic exposure assessment: an application of the PATH systematic observation method to retail workers. Postures, Activities, Tools and Handling. *Int J Occup Environ Health* 1999; 5(2): 79–87.

Ricci MG, De Marco F, Occhipinti E. Criteria for the health surveillance of workers exposed to repetitive movements. *Ergonomics* 1998; 41(9): 1357–1363.

Saldana N, Herrin GD, Armstrong TJ, Franzblau A. A computerized method for assessment of musculoskeletal discomfort in the workforce: a tool for surveillance. *Ergonomics* 1994; 37(6): 1097–1112.

Silverstein BA, Stetson DS, Keyserling WM, Fine LJ. Work-related musculoskeletal disorders: comparison of data sources for surveillance. *Am J Indust Med* 1997; 31: 600–608.

Warren N, Dillon C, Morse T, Hall C, Warren A. Biomechanical, psychosocial, and organizational risk factors for WRMSD: population-based estimates from the Connecticut upper-extremity surveillance project (CUSP) [In Process Citation]. *J Occup Health Psychol* 2000; 5(1): 164–181.

Chapter 10

Case definition for upper limb disorders

Francesco Violante, Lucia Isolani and Giovanni Battista Raffi

10.1 INTRODUCTION

In the surveillance of workers exposed to the risk of developing upper limb disorders, appropriate case definition plays an important role. In fact not only in the epidemiological field, but also in the clinical practice, appropriate case definition is essential in order to draw any valid generalized conclusion, not only for single patients.

In this chapter some general concepts regarding an illness case definition, diagnostic process and critical epidemiological evaluations for the surveillance of upper limb disorders in working populations are examined, and some case definitions of upper limb disorders potentially useful in the surveillance of workers at risk are proposed and discussed. In this context, when we want to compare risk factors and health outcome, they must both be clearly defined.

Different case definitions will affect directly the prevalence or incidence rate of the disease being studied, explaining in part different results in similar settings.

10.2 Disease and Medical Diagnosis

A disease can be defined as 'an impairment of the normal state of the living body or of one of its parts that interrupts or modifies the performance of the vital functions and is a response to environmental factors, to specific infective agents, to inherent defects of the organism or to combinations of these factors' (Merriam Webster's Medical Desk Dictionary, 1996).

Thus, illness is a condition characterized by a certain duration, by its evolution (it may improve or worsen) and by a certain degree of impairment in the person affected.

On the other hand (from a therapeutic perspective) an illness can be defined as an abnormal condition for which a person would usually seek medical treatment or for which there is consensus among doctors that medical treatment (if available) should be offered to the person (even though it is not requested).

However, medical diagnoses have specific components, and follow a relatively codified format, in which these components are elaborated, usually not quantitatively, by clinical opinion.

The process of clinical reasoning is poorly understood, but it is based on factors such as experience and learning, inductive and deductive reasoning, interpretation of evidence that itself varies in reproducibility and validity and intuition that is often difficult to define (Goldman 1998).

Typical components of a medical diagnosis are:

- History/Interview (Symptoms)
- Physical examination findings (Signs)
- Diagnostic procedure findings

Although it may be just a non-verified abstraction, some doctors tend to consider patients with specific symptoms as a set, those whose physical examination is pathologic as a subset of the former and those whose diagnostic procedures are also pathologic as a subset of the latter, as shown in Figure 10.1.

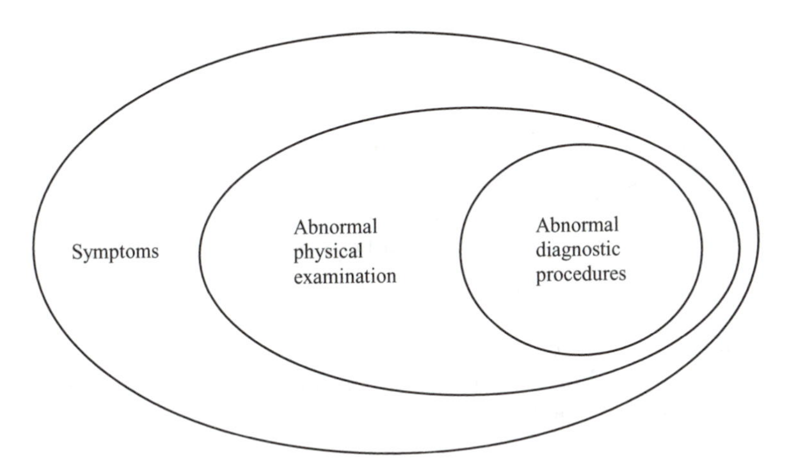

Figure 10.1 A theoretical assumption regarding relationships among symptoms, signs and diagnostic procedures.

Behind this concept, there is the assumption that symptoms on one hand are compatible with different pathologic entities, on the other hand they may stem not from the illness but from other individual characteristics: physical examination and diagnostic procedure would gradually select subjects identifying finally the really ill ones.

Subjectivity of symptoms has been criticized by those who think that other factors (e.g. 'workers compensation' economic gains or other types of gains) can induce over reporting of symptoms: regarding this, we must not forget that in working environments other factors are also relevant (e.g. job security) and they can induce an opposite effect.

Often instead, symptoms, signs and diagnostic procedures define independent sets, with a variable overlap, depending on the specific case as shown in Figure 10.2.

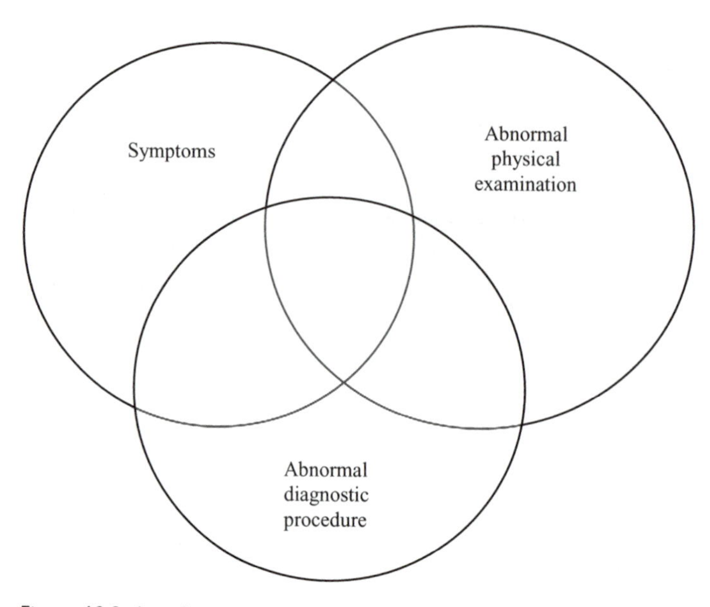

Figure 10.2 A realistic assumption regarding the relationship between symptoms, signs and diagnostic procedures.

As we will discuss later, data currently available on symptoms, signs and electro diagnostic procedures regarding carpal tunnel syndrome are consistent with this type of relationship.

10.3 CASE DEFINITION OF WORK RELATED UPPER LIMB DISORDERS

Current controversy surrounding occupational risk factors and upper limb disorders stems, in part, from differences in health outcome reported by different studies, which may be attributable in part to the different case definition used.

Some frequent work related upper limb disorders which will be further discussed are carpal tunnel syndrome and tendinitis: for each condition, a case definition useful in the surveillance of work related upper limb disorders is reported and discussed.

10.4 CARPAL TUNNEL SYNDROME

Among the work related upper limb disorders, carpal tunnel syndrome is the most studied and the one for which diagnostic criteria, to be used in epidemiological studies, originating from the consensus of many researchers, has been suggested.

Carpal tunnel syndrome is also a good model for understanding how symptoms, results of physical examinations and diagnostic procedures, overlap only in part, and how different results of published studies can be explained in part by different case definitions applied. Carpal tunnel syndrome is a neuropathy caused by the compression of the median nerve in its passage inside the carpal tunnel (Phalen 1966; Harrington et al. 1988). Generally the diagnosis of carpal tunnel syndrome (Anonymous 1985) hinges on the following:

- Symptoms: numbness, tingling or pain in the median nerve distribution (1st, 2nd and 3rd finger).
- Physical examination: positive Tinel's or Phalen's test, other tests (two points discrimination, vibrometry).
- Diagnostic procedures: electrodiagnostic studies indicative of nerve conduction block in the median nerve at wrist (prolonged sensory and motor conduction time, slowed sensory or motor conduction velocity).

Carpal tunnel syndrome has been studied extensively in working populations: its prevalence has been reported between 0.6 and 61% (Hagberg et al. 1992); less data are available on its incidence.

What is particularly striking is the wide range in the prevalence values, which varies about tenfold: at least in part this variability can be explained by the different case definition applied in different studies.

In a recent review of the literature (Bernard 1997) on the relationship between work and musculoskeletal illness, NIOSH identified several studies on different health outcome with quite restrictive inclusion criteria for the definitions of risk factors and health outcome: of the five studies on carpal tunnel syndrome which met the four inclusion criteria two out of three did not use (in the case definition) electrodiagnostic tests.

Some authors consider the electrodiagnostic studies to be the 'gold standard' for the diagnosis of carpal tunnel syndrome (Nathan et al. 1993), because they disclose the physiopathological process behind the illness and are also an objective test.

Nevertheless electrodiagnostic tests highlight a conduction block, whereas tingling, the typical symptom of the carpal tunnel syndrome, is the expression of an independent discharge of the nerve. Moreover, several studies have shown that among working populations the number of people with impaired results on electrodiagnostic tests exceeds the number of people with symptoms of carpal tunnel syndrome.

This finding is consistent with our experience: in a study on assembly line workers, 32 subjects were examined twice after about 24 months: in this group, out of seven hands with impaired electrodiagnostic examination, but without symptoms, only two developed symptoms, whereas most of the hands, with impaired electrodiagnostic parameters, became normal.

In the use of electrodiagnostic tests in the epidemiology of carpal tunnel syndrome another problem is the type of normal values against which measurements should be compared (Werner et al. 1997; Rempel et al. 1998).

In the clinical practice it is usually sufficient to use the value derived from a rather small set of normal subjects, which can be inappropriate for epidemiological studies in the working field.

For instance, in our experience with a group of meat processing workers, workers who did not present paresthesias in the median nerve distribution in their hands had nerve conduction values (Wrist Sensory Latency – WSL; Sensory Conduction Velocity – SCV; Wrist Motor Latency – WMC; Motor Conduction Velocity – MCV) intermediary between those of a group of normal subjects and of carpal tunnel syndrome patients described in a classical study by Kimura (Kimura, 1979), as shown in Table 10.1.

Table 10.1 Nerve conduction values in different groups of subjects.

	WSL (m/s) *Wrist* *Mean (sd)*	*SCV (m/s)* *Palm/Wrist* *Mean (sd)*	*WML (m/s)* *Wrist* *Mean (sd)*	*MCV (m/s)* *Palm/Wrist* *Mean (sd)*
Kimura, normal subjects	2.82 (0.28)	57.3 (6.9)	3.60 (0.4)	49.0 (5.7)
Asymptomatic hands of meat industry workers	3.20 (0.4)	48.2 (8.9)	4.30 (0.6)	34.4 (9.3)
Symptomatic hands of meat industry workers	3.50 (0.6)	41.8 (11.3)	4.80 (0.8)	29.5 (9.0)
Kimura, mild CTS patients	3.70 (0.59)	38.5 (7.5)	4.94 (1.0)	28.2 (7.5)

10.5 TENDINITIS

Tendinitis in different segments of the upper limb (hand wrist, elbow, shoulder) has been reported in association with different risk factors such as forceful or repetitive exertion in various working populations.

Regardless of the district affected, upper limb tendinitis shares some common features such as:

- Symptoms: recurrent pain in the tendon area, which increases during movements.

- Physical examination: pain increases during resisted movements or passive stretching of the tendons, limited range of motion, swelling or ganglia may be present.

Diagnosis of tendinitis, in most of the studies on the relationship between this illness and working activity, was established through the use of symptoms and physical examination.

In the last few years, in clinical practice ultrasonography has proved to be the elective means to study acute inflammation or to diagnose partial tendon fractures. Its usefulness can be compared to Computed Tomography or Magnetic Resonance Imaging.

Compared to CT and MRI, ultrasonography has the advantage of having no contraindications, it is fast and cheap to perform, it can be done with portable equipment and it can examine tendons dynamically: all these characteristics would make it a very useful diagnostic instrument for epidemiological studies, if in comparison with currently accepted diagnostic methods (symptoms and physical examinations) it gives indications in this direction.

Although much has been written about ultrasound of the hand and wrist, its role in the diagnosis of overuse tendinitis–peritendinitis is not clear at this time.

The scientific literature shows that tendons are easy to visualize using ultrasound (Van Holsbeeck and Introcaso 1990; Bouffard et al. 1993; Fornage and Rifkin 1988; Crass et al. 1988).

Examination of wrist tendons requires:

- use of high-frequency transducers (7.5–10 MHz) with a generous amount of sonographic gel and slight pressure on the transducer;
- the transducer probe must be accurately placed parallel to the tendon axis for longitudinal views and perpendicular for transverse scanning. This is because tendons are strongly anisotropic structures: they display different patterns depending on the angle of incidence of the sound beam. The greatest reflectivity is observed with the beam perpendicular to the surface of the tendon. So a tendon abnormality should be shown in both the transverse and longitudinal planes (Bouffard 1993);
- dynamic real-time imaging is needed to examine tendons throughout their physiological range of motion (every finding seen is part of tendon if it moves with it);
- comparative images with asymptomatic wrist, if it exists, is needed.

Sonographic findings of acute tenosynovitis are (Van Holsbeeck and Introcaso 1992; Bouffard et al. 1993; Fornage and Rifkin, 1988):

- tendon thickening;
- tendon hypoechogenicity (due to oedema, causing widening of the interfibrillar spaces);

- non-homogeneous hypoechogenicity of tendon (due to small tears of the fascicles and to foci of necrosis);
- thickening of tendon sheaths and presence of anechoic halo around the tendon on transverse images (due to an increased fluid within the synovial sheath).

Chronic tendinitis is a condition in which signs of trauma such as oedema, synovitis, and focal anatomic derangement are often hidden by the repair processes such as scarring (Lee 1996).

Hence, chronic tendinitis should be detected by findings similar but less marked to those of acute tenosynovitis. (Bouffard et al. 1993) such as:

- non-homogeneity of tendineous bundle echo texture: it is seen by the presence of both hypoechoic or anechoic images (due to deposition of mucoid material in clefts and micro ruptures of tendon fascicles) and hyperechoic linear striations (due to fibrosis); presence of hyperechoic areas with or without acoustic areas;
- thickening of tendon sheath: in normal tendons synovial sheaths are not well visualized. In chronic inflammatory wrist tenosynovitis the tendon sheath is distended with hypertrophic synovial tissue shown as homogeneous low level echoes (Van Holsbeeck and Introcaso 1992);
- presence of abnormal images associated with the tendon (calcifications, fluid within the synovial sheath).

Based on scientific literature review and on our own experience we worked out the protocol for sonographic diagnosis of tendinitis in the wrist described below.

10.6 A TENTATIVE PROTOCOL FOR ULTRASONOGRAPHIC DIAGNOSIS OF HAND WRIST TENDINITIS

Tendinitis of the digitorum flexors of the wrist can be recognized by the following ultrasound findings. Every one corresponds to a numeric value and final score has to be valuated as follows:

- 0–2: negative for tendinitis
- 3: doubt for tendinitis
- $\geq$4: positive for tendinitis

(1) Non-homogeneity of tendineous bundle echotexture

This is seen by the presence of hypoechoic or anechoic images, with irregular shape and various dimensions, isolated or partially confluent, together with hyperechoic linear striations: these remarks cause the disappearance of the normal fibrillar texture of tendon. The examination has to be performed in transverse and longitudinal orientations.

Scoring:

2 ☐ Definitively present
1 ☐ Doubtfully present
0 ☐ Absent

(2) Thickening of tendon sheath

This is detected by the presence of a marginal band, hyperechoic compared to tendon echogenicity. The hyperechoic band should be clearly distinguishable and should be at least 1 mm thick. The examination has to be performed in transverse and longitudinal orientations.

Scoring:

2 ☐ Presence in more than one tendon
1 ☐ Presence in one tendon
0 ☐ Absence

(3) Intratendineous calcifications

This is detected by the presence of one or more hyperechoic areas, with irregular shape and various dimensions, with or without acoustic shadow.

Scoring:

3 ☐ Presence in more than one tendon
2 ☐ Presence in one tendon
0 ☐ Absence

(4) Increased fluid within the synovial sheath

This is seen as an anechoic halo around the tendon on transverse images. On longitudinal views the tendon is bounded on both sides by anechoic lines. This anechoic image should be at least 1 mm thick.

Scoring:

3 ☐ Presence in more than one tendon
2 ☐ Presence in one tendon
0 ☐ Absence

10.7 REFERENCES

Anonymous. Diagnosis of the carpal tunnel syndrome. (Editorial). *The Lancet* 1985; 1: 854–855.

Bernard BP. *Musculoskeletal Disorders and Workplace Factors.* DHHS (NIOSH). Publication No. 97-141, 1997.

Bouffard JA, Eyler WR, Introcaso JH. Sonography of tendons. *Ultrasound Quarterly* 1993; 11(4): 259–286.

Crass JR, Van de Vegte GL, Harkavy LA. Tendon Echogenicity: ex vivo study. *Radiology* 1988; 167: 499–501.

Fornage BD, Rifkin MD. Ultrasound examination of tendons. *Radiol Clin North Am* 1988; 26: 87–107.

Goldman L. Quantitative aspects of clinical reasoning. In *Harrison's Principles of Internal Medicine*. McGraw-Hill, New York, 1998.

Hagberg M, Morgenstern H, Kelsh M. Impact of occupations and job tasks on the prevalence of carpal tunnel syndrome. *Scand J Work Environ Health* 1992; 18: 337–345.

Harrington JM, Carter JT, Birrel L, Gompertz D. Surveillance case definitions for work related upper limb pain syndromes. *Occup Environ Med* 1988; 55: 264–271.

Kimura J. The carpal tunnel syndrome. Localization of conduction abnormalities within the distal segment of the median nerve. *Brain* 1979; 102: 619–635.

Lee D. Wrist ultrasound unveils anatomic complexities. *Joint Imaging* 1996: 19–21.

Nathan P, Keniston RC, Meadows KD, Lockwood RS. Predictive value of nerve conduction measurement at the carpal tunnel syndrome. *Muscle and Nerve* 1993; 116: 1377–1382.

Phalen GS. The carpal tunnel syndrome. Seventeen years' experience in diagnosis and treatment of six hundred and fifty-four hands. *J Bone Joint Surg Am* 1966; 48(2): 211–228.

Rempel D, Evanoff B, Amadio PC, De Krom M, Franklin G, Franzblau A, Gray R, Gerr F, Hagberg, Hales T, Katz JN, Pransky G. Consensus criteria for the classification of carpal tunnel syndrome in epidemiologic studies. *Am J Pub Health* 1998; 88(10): 1447–1451.

Van Holsbeeck M, Introcaso JH. Musculoskeletal ultrasonography. *Radiol Clin North Am* 1992; 30(5): 907–925.

Van Holsbeeck M, Introcaso JH. Sonography of tendons. *Musculoskeletal Ultrasound.* Mosby-Year Book, Chicago, 1990.

Werner RA, Franzblau A, Albers JW, Buchele H, Armstrong TJ. Use of screening nerve conduction studies for predicting future carpal tunnel syndrome. *Occ Env Med* 1997; 54: 96–100.

Chapter 11

Back Risk Factors: An Overview

Stover H. Snook

11.1 INTRODUCTION

Low back pain and low back disability are not the same. They are certainly related, but they are very different (Allan and Waddell 1989). Low back pain is the subjective perception of pain in the lower back, buttocks, or leg (sciatica). There is no objective measure of pain; it can only be measured subjectively through self-report (pain drawings, pain analogues, pain scales, pain words, pain diaries, etc.).

Low back disability is the lost work time or restricted duty that results from low back pain. Low back disability is obviously related to low back pain, but includes many other variables that influence disability (e.g. type of job, management policies and practices, personality differences, supervisor conflicts, domestic problems, etc.). There is a poor relationship between low back pain and low back disability (Von Korff et al. 1993; Linton 1985; Snook et al. 1998).

11.2 LOW BACK PAIN

11.2.1 History of Low Back Pain

Allan and Waddell (1989) and Waddell (1998) have provided an interesting history of low back pain. According to these authors, the earliest surviving surgical text is the Edwin Smith papyrus from about 1500 BC. This text describes an episode of acute back strain, including the examination and diagnosis. When the treatment is described, it says 'Thou shouldst place him prostrate on his back; thou shouldst make for him . . .' And at this point, the scribe stopped copying, he subsequently died, and he and his papyrus were buried in a tomb near Thebes for nearly 3500 years.

In 1862, it was found by grave robbers, and sold to Edwin Smith. Hippocrates, about 400 BC, spoke about back pain, although it was not regarded as a major medical problem. Hippocrates also spoke about sciatica, the pain and numbness in the buttocks and down the leg into the foot. He described it as affecting men between 40 and 60, and usually lasting about 40 days. It is interesting that radiation of pain to the foot was regarded as a good prognostic sign, but if it stayed in the hip, it was dreaded. That was rather insightful for the time. Today, it is known that 50% of sciatica resolves itself within two weeks; and 70%, within six weeks. (Weinstein 1998).

There was very little medical thought during the Dark Ages when patient care moved into the hands of the church. But back pain persisted in folk medicine. The Welsh called it 'shot of the elf', and the Germans called it 'witches shot'.

Modern medicine, beginning with the Renaissance, classified back pain as rheumatism. Back pain was never thought to be an injury, at least not until the industrial revolution and particularly the building of the railroads in the 1800s. Only then was back pain blamed on trauma both from riding and being jostled on those early trains, as well as the labour involved in laying the track.

Modern medical treatment for back pain is closely linked to the evolution of orthopaedics and the key orthopaedic principle of rest. The founder of modern orthopaedics back in 1874 was Hugh Owen Thomas, who came from a long line of Welsh bone setters. Bone setters had treated back pain by manipulation and mobilization. They did this in the context of everyday life and their clients continued everyday activities. Orthopaedic practitioners moved back pain into a medical context. Back pain was now a disease and the sufferer was a patient. Orthopaedic treatment by rest, particularly by bed rest, removed the patient from everyday life. Today, it is known that bed rest is ineffective for low back pain and sciatica, and can actually be harmful.

Eighteen years later, in 1892, Andrew Still, a physician from the Mid-West, founded the American School of Osteopathy a reaction to orthodox medicine and the indiscriminate use of drugs. Dr Still combined the ancient principles of holistic medicine with manipulation from the bone setters.

Three years after that, in 1895, Wilhelm Roentgen developed X-rays, and for the first time one could view a living spine and every incidental X-ray finding was used to explain both back pain and sciatica. Today we know that there is little relationship between X-ray findings and low back pain, between cat scans and low back pain, and between MRIs and low back pain.

One year later, in 1896, Daniel David Palmer founded the first school of chiropractice in Iowa, based on manipulation of subluxations (or misalignment), but also upon what they called vitalism or an emphasis of the mind–body relationship and the patient's capacity for self-healing.

Physical therapy began in the U.S. during World War I, with the work of Mary MacMillan on maimed and disabled soldiers. This led to the formation of the American Physical Therapy Association in 1921, with a strong emphasis on exercise and rehabilitation.

Modern disc surgery began in 1932 at Massachusetts General Hospital, when neurosurgeon William Jason Mixter and orthopaedist Joseph S. Barr operated for the first time on a patient with the diagnosis of 'ruptured intervertebral disc'. By the 1950s, there was an explosion of disc surgery, which was closely linked to the growth of neurosurgery as a specialty. Today the rate of disc surgery in the U.S. is five times greater than in Great Britain. Disc surgery has been accused of leaving more tragic human wreckage in its wake than any other operation in history (DePalma and Rothman 1970).

Despite this long history, low back pain has not enjoyed the medical advances that are found in other areas of medicine. For example, we have seen dramatic reductions in small pox, polio, and tuberculosis – but there is absolutely no evidence that low back pain has declined over the years.

11.2.2 Prevalence of low back pain

At any given point in time (point prevalence), 15% to 20% of adults experience symptoms of low back pain (Frymoyer and Cats-Baril 1991). The one-month prevalence of low back pain is estimated between 35% and 37% (Papageorgiou et al. 1995). The one-year prevalence is approximately 50% (Lawrence et al. 1998), and the lifetime prevalence is estimated between 60% and 80% (Frymoyer and Cats-Baril 1991). The lifetime prevalence of true sciatica (true nerve root pain) is estimated between 3% and 5% (Waddell 1994, 1998).

11.2.3 Cause of low back pain

Up to 85% of low back pain has no definite etiology, and is classified as idiopathic or non-specific. (White and Gordon 1982; Spitzer et al. 1987). There are many theories, but no hard data. 'We currently cannot differentiate pain producing pathology from simple aging changes' (Videman and Battié 1996).

It is inappropriate to refer to all back symptoms occurring in the workplace as injuries (Battie 1992). Most back symptoms are of spontaneous and gradual onset, without a specific precipitating accident or unusual activity. Hall (1998) reported that 66.7% of non-compensated low back pain patients (with no requirement to demonstrate a cause) could not identify a precipitating event; as opposed to 9.8% of compensated patients who could not identify an event.

'The main suspected source of pain is pathology in the outer layers of the anulus and around the nerve roots' (Videman and Battie 1996). According to the theory of internal disc disruption, radial fissures extend from the nucleus pulposus to the innervated, outer third of the anulus fibrosus. It is thought that the fissures expose the nerve endings of the outer anulus to the noxious or inflammatory material of the nucleus (Bogduk 1991).

11.2.4 Risk Factors of Low Back Pain

Increasing age

The body changes with increasing age, and the intervertebral disc is one of the earliest parts of the body to change. For example, the direct blood supply to the disc is lost somewhere between age 15 and 20, making the disc the largest avascular structure in the body (Nachemson 1976). There is no evidence of anular tears before age 20. According to Miller et al. (1988), 7% of people in their 20s exhibit anular tears; 20%, in their 30s; 41%, in their 40s; 53%, in their 50s; 85%, in their 60s; and 92%, for people over 70.

The prevalence of back pain increases from early adult life to the late forties or early fifties and remains relatively constant thereafter, at least to the mid-sixties. In those who continue to have back pain it is likely to be more frequent or more constant with increasing age (Waddell 1994, 1998).

The symptoms also change with age (Rowe 1983). People in their twenties and early thirties usually suffer from sudden acute attacks of short duration. During the mid to late thirties, the pain often becomes more localized to one side or the other, and listing to one side becomes evident. There may also be some residual, mild backache between attacks. Radiation of pain into the buttocks, the thigh, and even down to the foot happens more frequently in the forties. And in the fifties, the pain often becomes more constant, but less severe (more of an arthritic type symptom).

Prior episode

The probability of an episode of low back pain is greater after the initial episode; four times greater according to Dillane et al. 1966. Studies consistently find a history of back symptoms to be an indicator of future risk. (Videman and Battié 1996).

Occupation

The literature regarding the relationship between physical demands and low back pain is contradictory (Hall et al. 1998; Waddell 1994). There is little evidence to support the belief that heavy manual work causes low back pain (Waddell 1998). 'Whether physical loading is important in the etiology of conditions underlying back symptoms, or whether it primarily serves to exacerbate symptoms from an existing condition, is unclear' (Videman and Battie, 1996). Bending and twisting alone, without heavy lifting, have been associated with low back pain (Andersson1997; Snook et al. 1998). This may explain why light and sedentary workers also develop low back pain.

Time of day

The back is more vulnerable during the early morning hours. The probability of sprain and strain disorders is greater between 6 a.m. and 11 a.m. than at any other time of day (Choi et al. 1995; Fathallah and Brogmus 1999). Bending stresses are three times greater in the early morning than they are later in the day (Adams et al. 1987). There is an increased risk of disc problems when bending forward in the early morning, primarily due to increased fluid in the discs at that time (Adams et al. 1987). Significant reductions in pain were demonstrated in a randomized, controlled trial of 60 subjects, simply by not having them bend forward during the first hour of the day (Snook et al. 1998).

Genetics

There is a familial predisposition toward lumbar disc pain. (Richardson et al. 1977; Matsui et al. 1998) An identical twin study showed that genetics was more important than occupational factors in determining disc degeneration (Battié et al. 1995).

Obesity

There is a steady increase in back pain prevalence with increasing obesity, but most strikingly in the highest 20% of body mass index (Deyo and Bass 1989).

Smoking

The risk of low back pain increases steadily with cumulative exposure and with degree of maximal daily exposure (Deyo and Bass 1989).

11.2.5 Treatment of low back pain

Low back pain can be very painful, but less than 1% of it is serious spinal disease, less than 1% is inflammatory disease, and less than 5% is true nerve root pain (and only a small portion of these cases ever need surgery). The rest of it is simple backache, i.e. non-specific low back pain (Waddell 1998).

There are many types of practitioners who treat low back pain general practitioners, chiropractors, orthopaedic surgeons, physiatrists, neurosurgeons, neurologists, osteopaths, physical therapists, etc. However, 61% of adults with acute severe low back pain did not seek any health care during their most recent episode of pain (Carey et al. 1996).

The treatment varies greatly, depending on whom you go to from bed rest to exercise to manipulation to medication to heat to cold to traction to acupuncture to injections to surgery, etc. However, it doesn't make much difference what type of practitioner is visited, or what kind of treatment is given. The results are about the same (Waddell 1987; Carey et al. 1995; Cherkin et al. 1998; Shekelle 1998). There are differences in costs, in patient satisfaction, and in the number of visits; but the results are about the same. Health care providers themselves, with greater knowledge about the problem, have the same amount of low back pain as the general population (Scholey and Hair 1989; Boos 1995).

In the United States, medical care for low back pain is overspecialized, overinvasive, and overexpensive, whereas in the United Kingdom, National Health Service care for back pain is underfunded, too little, and too late. Yet, there is little difference in clinical outcome or social impact of treatment in the two countries. (Waddell 1996) 80% of the people get better no matter what we do, and 20% of the people do not get better no matter what we do (Deyo 1996).

According to Waddell (1996), 'Medical care has not solved the low back pain problem, and may even be reinforcing and exacerbating the problem ... Back pain is a 20th century health care disaster.'

Treatment is difficult to evaluate because of two powerful factors. The first factor is the natural course or the natural history of the disorder. The primary characteristic of low back pain is intermittency. It comes and goes. From a scientific point of view, it must be shown that the treatment produces a faster and better outcome than no treatment at all. The second factor is the placebo effect. The placebo effect refers to the positive results from just being treated, regardless of the type of treatment. It can be very strong and account for 55% to 60% of a positive outcome. (Weinstein 1999). A practitioner is happy for any kind of a positive outcome, but a scientist would like to know how much of the outcome is due to the treatment, and how much is due to the placebo.

11.2.6 Prognosis of low back pain

Most patients improve considerably during the first four weeks (Von Korff and Saunders 1996). Weinstein (1998) reports that 50% of sciatica resolves itself in two weeks; and 70%, within six weeks. Nachemson (1992) observes that 90% of patients with single episodes return to work within six weeks.

However, low back pain among primary care patients typically runs a recurrent course characterized by variation and change, rather than an acute, self-limiting course. For those experiencing low back pain for the first time, approximately 70% will still report pain one year later (Von Korff et al. 1993; Wahlgren et al. 1997). The recurrence rate of low back pain is very high (Nachemson 1992; Rossignol et al. 1992; MacDonald et al. 1997). Miedema et al (1998) found that 28% of low back pain becomes chronic after seven years.

11.2.7 Prevention of low back pain

There is no evidence that low back pain has decreased during the last 20 years (Burton 1997). However, the disability from low back pain increased dramatically between the 1950s and 1970s without an increase in low back pain (Waddell 1987). One study showed that low back disability from 1957 to 1976 increased at a rate that was 14 times greater than the rate of population growth (Frymoyer and Durett 1997). The major problem in industry today is low back disability, not low back pain.

There are a growing number of investigators who believe that efforts at preventing low back pain are futile; that low back pain is an unavoidable consequence of life that will afflict most people at some point in their lives, regardless of the nature of their work (Rowe 1983; Frank et al. 1996a; Snook 1989; Spitzer 1993). These investigators believe that programmes aimed at preventing low back disability are likely to be much more effective and less costly. As Frymoyer and Cats-Baril (1991) point out, 'The future challenge, if costs are to be controlled, appears to lie squarely with prevention and optimum management of disability, rather than perpetrating a myth that low back pain is a serious health disorder.'

11.3 LOW BACK DISABILITY

11.3.1 History of low back disability

There is little mention of low back disability in ancient times. The first report of work-related back pain was by Ramazzini, the founder of occupational medicine, in 1705. The first report of low back disability was on the railways in the U.K. A Lancet Commission in 1862 found that railway workers had more sickness than seaman, miners or labourers. Lumbago was one of the most common causes. (Waddell 1998; Allan and Waddell 1989).

'Railway spine' became an increasing problem between 1860 and 1880, and the first reports of prolonged low back disability were in the 1880s and 1890s. Low back disability became a more widespread problem during the first two decades of the 20th century, and has increased dramatically since World War II. There was very little mention of low back disability in women. Only recently has sexual equality allowed women to have low back disabiliity (Waddell 1998; Allan and Waddell 1989).

11.3.2 Incidence of low back disability

Fordyce (1995) reports incident rates for low back disability in Table 11.1.

Table 11.1 Low back pain disability rates by country.

	Incidence rate (annual percentage)	*Days absent (per patient per year)*
United States	2	9
Canada	2	20
Great Britain	2	30
West Germany	4	10
The Netherlands	4	25
Sweden	8	40

A study by Honeyman and Jacobs (1996) illustrates the importance of culture in reporting low back disability. They found that nearly half of the adults in a small Australian aboriginal community experienced long-term private spinal pain. However, because of their cultural beliefs, the aboriginals did not commonly make their pain public.

In 1992, 18% of all Liberty Mutual workers' compensation claims were for low back pain; 48% of these claims involved disability (days away from work) (Webster 1996). Substantial reductions in length of disability have been reported in the U.S.A. during the 1990s, probably due to a combination of injury prevention programmes, disability management, managed care, decreased medical inflation, economic expansion and reforms in workers' compensation laws (Hashemi et al. 1998).

11.3.3 Cost of back disability

In the United States, the total costs of low back pain are estimated between 50 and 100 billion dollars per year (Frymoyer and Cats-Baril 1991). 'Although new technologies ... have driven some of these costs, the single major factor has been the increasing numbers of people disabled by low back disorders, particularly those with workers compensation' (Frymoyer 1993).

In 1992, 30% of Liberty Mutual workers' compensation claims costs were for low back disorders. Of those costs, 36% were for medical costs; 60%, for indemnity (disability) costs (Webster 1996). The mean cost per low back pain claim was $7770; median, $518. The mean cost for all workers' compensation claims was $4590; median, $262.

Only 12.4% of Liberty Mutual low back claims had a length of disability greater than three months, but these claims accounted for 87% of the total compensation costs (Hashemi et al. 1997). Substantial reductions in costs associated with long disability has been reported in the U.S. during the 1990s (Hashemi et al. 1998). The top three sources of medical costs in the U.S.A. are diagnostic procedures (MRIs, CT scans, etc.) (25%), surgery (21%), and physical therapy (20%) (Williams et al. 1998).

11.3.4 Cause of low back disability

Low back pain

It is obvious that, in some cases, low back pain results in disability. However, contrary to popular opinion, the majority of people with low back pain continue to work (2% disability per year vs. 50% pain per year).

Compensation insurance

Low back disability is closely related to the development of compensation insurance. However, according to Allan and Waddell (1989), 'it is wrong to infer that disability is caused by compensation. Indeed the converse is true: legislation for compensation was only passed after a need was recognised.'

The best available data, as reviewed by Loeser et al. (1995) suggests that a 10% increase in workers' compensation benefits produces a 1% to 11% increase in the number of claims, and a 2% to 11% increase in the average duration of claims. In addition, compensation patients have poorer results from back surgery, and do not respond as well to pain management and rehabilitation (Waddell 1998). However, Waddell reminds us that health care professionals benefit more from the compensation system than any back pain patient ever did.

Occupational factors (type of job)

Unlike low back pain, low back disability is closely related to the type of job performed. There is almost unanimous agreement that people with heavy manual jobs lose more time from work because of back pain than workers with lighter jobs. Lifting is more frequently associated with low back pain workers' compensation claims (49%) than any other task or movement (Snook et al. 1978); particularly lifting from the floor, lifting bulky objects, lifting heavy objects, and lifting frequently (Waddell 1998).

Bending, without lifting, is also a problem. Back disorders in an automobile assembly plant were associated with mild (20–45°) trunk flexion (odds ratio: 4.9), severe (>45°) trunk flexion (odds ratio: 5.7), and trunk twist or lateral bend (odds ratio: 5.9) (Punnett et al. 1991).

In urban transit operators, part time work was associated with a 2.7-fold decrease in low back pain workers' compensation claims, and heavy physical labor (cable car crews) was associated with a 3-fold increase (Krause et al. 1998b).

Psychosocial variables

'If the risk factors for low back disability are analyzed, it becomes clear that this is not a medical problem ... Factors that are important are dominantly psychosocial and include poor health habits, job dissatisfaction, less appealing work environments, poor ratings by supervisors, psychologic disturbances, compensatory injury, and histories of prior disability' (Frymoyer and Cats-Baril 1991).

In urban transit operators, low back pain workers' compensation claims were predicted by psychological job demands (odds ratio: 1.50), job dissatisfaction (odds ratio: 1.56), frequency of job problems (odds ratio: 1.52), low supervisor support (odds ratio: 1.30), and female gender (odds ratio: 1.50) (Krause et al. 1998b).

Health care providers

The following quotations by prominent physicians emphasize the role of health care providers in causing or prolonging low back disability:

> Indeed, one is drawn to the conclusion that low back disability may well be an iatrogenic disorder in many cases.
>
> (Hrudey 1991)

> Sadly, we must conclude that much low back disability is iatrogenic.
>
> (Allan and Waddell 1989)

> Low back disability is 'aided and abetted by the health care provision system in general and by doctors and physiotherapists in particular.'
>
> (Spitzer 1993)

> An initial specific diagnosis by the treating physician was 'highly predictive of chronic disability from back pain, especially in older workers.'
>
> (Abenhaim et al. 1995)

> ... early aggressive treatment in the acute phase (3–4 weeks) has the potential for iatrogenesis ... there are many who would do well without any treatment. Some of these patients are at risk for increased sick-role behavior and, consequently, a longer period of disability as a result of well-intended treatment programs that are too intensive for the expected prognosis at this stage ...
>
> (Frank et al. 1998)

> ... have medical professionals of all types become part of the problem, rather than the solution?
>
> (Frymoyer and Cats-Baril)

Management

Management often complains that the high costs of low back pain are the result of disgruntled workers, incompetent practitioners, and militant unions. However, management is also part of the problem by not responding well to low back pain when it does occur among workers. For example, only half of the workers on disability are contacted by their immediate supervisor, despite evidence that supervisor concern and contact is associated with reduced disability (Workers' Compensation Monitor 1998).

The Michigan Disability Prevention Study demonstrated that lower levels of disability are associated with management policies and practices, particularly safety diligence, safety training, and proactive return-to-work programmes (Hunt and Habeck 1993). The probability of getting off disability and returning to work after various durations of disability is illustrated in Tables 11.2 and 11.3.

Table 11.2 Probability of getting off disability after one year (Hashemi 1997).

Disabled for 1 month	62%
Disabled for 3 months	44%
Disabled for 6 months	28%
Disabled for 9 months	14%

Table 11.3 Probability of returning to work.

	McGill 1968	*Rosen 1986*
Off work for 6 months	50%	35–50%
Off work for 1 year	25%	10–25%
Off work for 2 years	Nil	2–3%

Labour unions

According to Spitzer (1993), the Chairman of the Quebec Task Force on Spinal Disorders, 'The Quebec Workmens' Compensation Board (CSST) to this day has failed to implement any clinical or organisational recommendations of the task force. It is generally acknowledged that this is largely because of strong opposition by organised labour. The unions never liked the key conclusion of the report:

> 'The best treatment for low back pain without radiation or objective clinical signs is work.'

Rigid work rules often prevent early return to work. Also detrimental are union referrals to 'friendly' physicians who prolong disability, and referrals to 'friendly' attorneys who press for lump sum settlements instead of rehabilitation (Snook 1987).

11.3.5 CONTROL OF LOW BACK DISABILITY

Although low back pain may not be preventable at the present time, the good news is that we know how to reduce the disability from low back pain. There is sufficient evidence in the literature to suggest an approach consisting of the following components:

Management commitment

Occupational safety and health literature stresses the importance of management commitment in any successful safety and health effort (U.S. General Accounting Office 1997). Specific ways in which management commitment can be demonstrated include establishing goals, evaluating results, assigning staff, providing staff time, making resources available, and providing written communication to all employees regarding the importance of the programme in reducing low back disability. The intent is to create a corporate culture with a positive and supportive attitude regarding employees with low back pain. Trust building and employee advocacy are important ingredients.

Supervisor training

Supervisors must be trained in the true nature of non-specific low back pain, i.e. that low back pain is a disorder of unknown cause that happens to practically everyone, that low back pain usually develops gradually and insidiously without an unusual or strenuous activity, that low back pain may recur frequently, and that low back pain does not respond well to treatment but usually resolves itself within a few days or weeks (Snook 1989).

The best way for a supervisor to respond to a disorder of this type is to show some concern for the needs of the employee, to avoid making judgements and setting up adversary relationships, to encourage the employee to seek appropriate medical treatment, and to consider adapting the workplace (or modifying the job) so the employee may continue working on the job. This was the approach used by Fitzler and Berger (1982, 1983) at American Biltrite, Inc., where low back pain compensation costs were reduced by 90%.

Communication with employees

> Workers have come to believe that any back pain they may experience is a likely product of their work, and societal forces have tended to reinforce this beliefAttribution in particular is an important factor for compliance with intervention strategies, and it has been found that a simple educational program that detunes perception of attribution is capable of creating a positive shift in beliefs with a concomitant reduction in extended sickness absence.
>
> (Burton, 1997)

As with supervisors, employees should also be trained in the true nature of low back pain, before an episode occurs. If a disabling episode does occur, the supervisor should call or visit the employee within the first day or two, not to intimidate the employee, but to let him/her know that the company is supportive. The message should be: 'You are a vital part of our team, your work is important, and your job is waiting for you.' This was the approach used by Wood (1987) with staff in a geriatric hospital, where long-term low back disability claims were reduced from 7.1% to 1.7%.

Coordination with organized labour

Traditionally, labour unions are opposed to early return to work after a disabling episode of low back pain. It is thought that the employee is entitled to time off for even a minor episode, despite sound medical evidence that activity and work will hasten healing (Derebery and Tullis 1983; Spitzer 1993; Nachemson 1983; Malmivaara et al. 1995). Organized labour must be involved in the planning and execution of a disability prevention programme. There must be agreement between labour and management on what constitutes the best interests of the employee.

Workplace redesign (ergonomics)

Workplaces must be designed for people with low back pain as well as for people without low back pain. Good workplace design will permit employees with low back pain to remain on the job, or to return to the job sooner. Hadler (1997a) suggests that ergonomics 'turn from the quest to prevent back 'injuries' to the quest for enhancing the ability of a person with a backache to cope.' We should 'provide workplaces that are comfortable when we are well and accommodating when we are ill' (Hadler 1997b). The involvement of employees in any kind of job redesign is essential.

Management is often reluctant to redesign jobs because of the cost involved. However, committing capital to redesign jobs can often be a wise business investment. Decreases in compensation costs and increases in worker performance will return the cost of the initial investment over time. Determining the 'payback period' will help convince management of the cost effectiveness of redesigning jobs.

Proactive return to work programme

Several studies have shown that the longer an employee is disabled by low back pain, the lower is the probability of getting off disability and returning to work by a certain time (Hashemi et al. 1997; Andersson et al. 1983; McGill. 1968). These studies emphasize the importance of providing modified, alternative or part-time work as a means of returning the disabled employee to the job as quickly as possible. A supportive workplace response to low back pain needs to start when the pain is first reported; an individualized and accommodative approach to return to work should follow promptly. (Frank et al. 1998). A recent review of the literature concluded that modified work programmes reduce the number of lost work days by 50% (Krause et al. 1998a).

A proactive return to work programme is a supportive, company-based intervention for personally assisting the disabled employee from the beginning of the episode to its positive resolution. In a proactive programme, the actions and responsibilities of individuals within the company and external providers are spelled out and related to the goal of resumption of employment. In the Michigan Disability Prevention Study, companies reporting a 10% greater level of achievement on the proactive return to work variable demonstrated a significant 13.6% lower rate of lost workday cases ($p < 0.01$) (Hunt and Habeck 1993).

Selection of appropriate health care providers

Management should select health care providers who diagnose and treat low back pain according to accepted guidelines (Agency for Health Care Policy and Research 1994; Clinical Standards Advisory Group 1994; American College of Occupational and Environmental Medicine 1997; Royal College of General Practitioners 1996). Non-specific low back pain is benign. It should be treated and managed by primary care providers, and not referred to specialists without clear indications. According to the eminent orthopaedic surgeon, Gordon Waddell (1996), 'Orthopedic surgeons in particular are the wrong specialists to provide or control health care for non-specific low back pain ... physical therapy for non-specific low back pain should change from symptomatic methods, which have been shown to be ineffective, to early activation and restoration of function ...' (Waddell 1996).

The making of a 'specific' diagnosis at the beginning of a compensated episode carries the message that the condition is serious and that a 'specific' clinical procedure must be carried out. One consequence of this 'labelling effect' is to investigate and treat the lesion suspected of being the cause of the pain, rather than focusing on the functional recovery. 'This situation encourages patients to believe there is a cure for their problem when it is known that only a small number will respond to a specific therapy' (Abenhaim et al. 1995). Linton (1998) refers to 'secondary prevention' as quality pain management at the primary care level. 'The idea is to provide better multidimensional care, a little earlier, and with better coordination with other agents (e.g. the workplace, insurance companies, and authorities).'

11.4 SUMMARY

At the present time, the prevention and treatment of low back pain is ineffective. However, the problem in industry is not low back pain; it is the disability that results from low back pain. Fortunately, low back disability can be reduced if there is the will and motivation to apply the knowledge that presently exists.

Management must assume a more active role; they are the principal defence in the battle to reduce low back disability. Management must recognize that the primary responsibility for reducing low back disability belongs to them not to the medical establishment, not to the insurance company, not to the Workers' Compensation Board, but to themselves.

Health care providers must also recognize that they have an important, but limited, role in reducing low back disability. Health care providers must first attend to the patient with a correct and objective diagnosis, appropriate medication to reduce symptoms, referral to specialists only if clearly indicated, patient education and counselling, and early activation and return to work. An equally important role for the health care provider is to work with management in developing training programmes, ergonomics programmes, and policies and practices that have been shown to be effective in reducing low back disability.

The Michigan Disability Prevention Study concluded that 'disability can be managed; and those who do it well can expect to be rewarded with lower disability costs, more satisfied workers, greater productivity and, ultimately, higher profits' (Hunt and Habeck 1993).

11.5 REFERENCES

Abenhaim L, Rossignol M, Gobeille D, Bonvalot Y, Fines P, Scott S. The prognostic consequences in the making of the initial medical diagnosis of work-related back injuries. *Spine* 1995; 20, 791–795.

Adams MA, Dolan P, Hutton WC. Diurnal variations in the stresses on the lumbar spine. *Spine* 1987; 12, 130–137.

Agency for Health Care Policy and Research. Acute Low Back Problems in Adults. Clinical Practice Guideline No. 14. AHCPR Publication No. 95-0642. Public Health Service, U.S. Department of Health and Human Services, Rockville, MD, 1994.

Allan DB, Waddell G. An historical perspective on low back pain and disability. *Acta Orthopaedica Scandinavica* 1989; 60(Suppl. No. 234), 1—23.

American College of Occupational and Environmental Medicine. Low back complaints, Chapter 14, *Occupational Medicine Practice Guidelines*, JS Harris, ed. OEM Press, Beverly, MA, 1997.

Andersson GBJ, The epidemiology of spinal disorders. In *The Adult Spine: Principles and Practice*, JW Frymoyer, ed. Lippincott-Raven, Philadelphia, 1997.

Andersson GBJ, Svensson HO, Odén A. The intensity of work recovery in low back pain. *Spine* 1983; 8, 880–884.

Battié, MC. Minimizing the impact of back pain: workplace strategies. *Seminars in Spine Surgery* 1992; 4, 20–28.

Battié, MC, Videman T, Gibbons LE, Fisher LD, Manninen H, Gill K. Determinants of lumbar disc degeneration. *Spine* 1995; 20, 2601–2612.

Bogduk N. The lumbar disc and low back pain. *Neurosurgery Clinics of North America* 1991; 2, 791–806.

Boos N. The spines of the rich and famous. *The BackLetter* 1995; 10, 96.

Burton AK. Spine update: back injury and work loss. *Spine* 1997; 22, 2575–2580.

Carey TS, Evans AT, Hadler NM, Lieberman G, Kalsbeek WD, Jackman AM, Fryer JG, McNutt RA. Acute severe low back pain: a population-based study of prevalence and care-seeking. *Spine* 1996; 21, 339–344.

Carey TS, Garrett J, Jackman A, McLaughlin C, Fryer J, Smucker DR. The ourcomes and costs of care for acute low back pain among patients seen by primary care practitioners, chiropractors, and orthopedic surgeons. *New England Journal of Medicine* 1995; 333, 913–917.

Cherkin DC. Primary care research on low back pain: the state of the science. *Spine* 1998; 23, 1997–2002.

Cherkin DC, Deyo RA, Battie, M, Street J, Barlow W. A comparison of physical therapy, chiropractic manipulation, and provision of an educational booklet for the treatment of patients with low back pain. *New England Journal of Medicine* 1998; 339, 1021–1029.

Choi BCK, Levitsky M, Lloyd RD, Stones IM. *Analysis of 1990 sprain and strain injuries in Ontario*, Workplace Health & Safety Agency, Toronto, 1995: 1–27.

Clinical Standards Advisory Group. Report of a CSAG Committee on Back Pain. National Health Service, HMSO, London, 1994.

DePalma AF, Rothman RH. *The Intervertebral Disc*. WB Saunders Co., Philadelphia, 1970.

Derebery VJ, Tullis WH. Delayed recovery in the patient with a work compensable injury. *Journal of Occupational Medicine* 1983; 25, 829–835.

Deyo RA. Low back pain: a primary care challenge. *Spine* 1996; 24, 2926–2832.

Deyo RA, Bass JE. Lifestyle and low-back pain: the influence of smoking and obesity. *Spine* 1989; 14, 501–506.

Dillane JB, Fry J, Kalton G. Acute back syndrome: a study from general practice. *British Medical Journal* 1966; 2, 82–84.

Fathallah FA, Brogmus GE. Hourly trends in workers' compensation claims. *Ergonomics* 1999, 42, 196–207.

Fitzler SL, Berger RA. Attitudinal change: the Chelsea back program. *Occupational Health and Safety* 1982; 51, 24–26.

Fitzler SL, Berger RA. Chelsea back program: one year later. *Occupational Health and Safety* 1983; 52, 52–54.

Fordyce WE. The problem. In *Back Pain in the Workplace: Management of Disability in Nonspecific Conditions,* WE Fordyce, ed. International Association for the Study of Pain (IASP) Press, 1995: 5–9.

Frank JW, Kerr MS, Brooker AS, DeMaio SE, Maetzel A, Shannon HS, Sullivan TJ, Norman RW, Wells RP. Disability resulting from occupational low back pain: Part I: What do we know about primary prevention? A review of the scientific evidence on prevention before disability begins. *Spine* 1996a; 21, 2908–2917.

Frank JW, Brooker AS, DeMaio SE, Kerr MS, Maetzel A, Shannon HS, Sullivan TJ, Norman RW, Wells RP. Disability resulting from occupational low back pain: Part II: What do we know about secondary prevention? A review of the scientific evidence on prevention after disability begins. *Spine* 1996b; 21, 2918–2929.

Frank JW, Sinclair S, Hogg-Johnson S, Shannon H, Bombardier C, Beaton D, Cole D. Preventing disability from work-related low-back pain. *Canadian Medical Association Journal* 1998; 158, 1625–1631.

Frymoyer JW. An international challenge to the diagnosis and treatment of disorders of the lumbar spine. *Spine* 1993; 18, 2147–2152.

Frymoyer JW, Cats-Biril WL. An overview of the incidences and costs of low back pain. *Orthopedic Clinics of North America* 1991; 22, 263–271.

Frymoyer JW, Durett CL. The economics of spinal disorders. In *The Adult Spine: Principles and Practice,* second edition, JW Frymoyer, ed. Lippincott-Raven, Philadelphia, 1997.

Hadler NM. Editorial: back pain in the workplace. *Spine* 1997a; 22, 935–940.

Hadler NM. Workers with disabling back pain. *New England Journal of Medicine* 1997a; 337, 341–343.

Hall H, McIntosh G, Wilson L, Melles T. Spontaneous onset of back pain. *Clinical Journal of Pain* 1998; 14, 129–133.

Hashemi L, Webster BS, Clancy EA, Volinn E. Length of disability and cost of workers' compensation low back pain claims. *Journal of Occupational and Environmental Medicine* 1997; 39, 937–945.

Hashemi L, Webster BS, Clancy EA. Trends in disability duration and cost of workers' compensation low back pain claims (1988–1996). *Journal of Occupational and Environmental Medicine* 1998; 40, 110–119.

Honeyman PT, Jacobs EA. Effects of culture on back pain in Australian aboriginals. *Spine* 1996; 21, 841–843.

Hrudey WP. Overdiagnosis and overtreatment of low back pain: long term effects. *Journal of Occupational Rehabilitation* 1991; 1, 303–311.

Hunt HA, Habeck RV. *The Michigan Disability Prevention Study.* W. E. Upjohn Institute for Employment Research, Kalamazoo, MI, 1993.

Krause N, Dasinger LK, Neuhauser F. Modified work and return to work: a review of the literature. *Journal of Occupational Rehabilitation* 1988a; 8, 113–139.

Krause N, Ragland DR, Fisher JM, Syme SL. Psychosocial job factors, physical workload, and incidence of work-related spinal injury: a 5-year prospective study of urban transit operators. *Spine* 1998b; 23, 2507–2516.

Lawrence RC, Helmick CG, Arnett FC, Deyo RA, Felson DT, Giannini EH, Heyse, SP, Hirsch R, Hochberg MC, Hunder GG, Liang MH, Pillemer SR, Steen VD, Wolfe F. Estimates of the prevalence of arthritis and selected musculoskeletal disorders in the United States. *Arthritis and Rheumatism* 1998; 41, 778–799.

Linton SJ. The relationship between activity and chronic back pain. *Pain* 1985; 21, 289–294.

Linton SJ. Editorial: the socioeconomic impact of chronic back pain: is anyone benefiting? *Pain* 1995; 75, 163–168.

Loeser JD, Henderlite SE, Conrad DA. Incentive effects of workers' compensation benefits: a literature synthesis. *Medical Care Research and Review* 1995; 52, 34–59.

MacDonald MJ, Sorock GS, Volinn E, Hashemi L, Clancy EA, Webster B. A descriptive study of recurrent low back pain claims. *Journal of Occupational and Environmental Medicine* 1997; 39, 35–43.

Matsui H, Kanamori M, Ishihara H, Yudoh K, Naruse Y, Tsuji H. Familial predisposition for lumbar degenerative disc disease. *Spine* 1998; 23, 1029–1034.

McGill CM, Industrial back problems: a control program. *Journal of Occupational Medicine* 1968; 10, 174–178.

Malmivaara A, Hakkinen U, Aro T, Heinrichs ML, Koskenniemi L, Kuosma E, Lappi S, Paloheimo R, Servo C, Vaaranen V, Hernberg S. The treatment of acute low back pain - bed rest, exercises, or ordinary activity? *New England Journal of Medicine* 1995; 332, 351–355.

Miedema HS, Chorus AMJ, Wevers CWJ, van der Linden S. Chronicity of back problems during working life. *Spine* 1998; 18, 2021–2029.

Miller JAA, Schmatz C, Schultz AB. Lumbar disc degeneration: corelation with age, sex, and spine level in 600 autopsy specimens. *Spine* 1988; 13, 173–178.

Nachemson A. The lumbar spine: an orthopaedic challenge. *Spine* 1976; 1, 59–71.

Nachemson A. Work for all. *Clinical Orthopaedics and Related Research* 1983; 179, 77–85.

Nachemson AL. Newest knowledge of low back pain: A critical look. *Clinical Orthopaedics and Related Research* 1992; 279, 8–20.

Papageorgiou AC, Croft PR, Ferry S, Jayson MIV, Silman AJ. Estimating the prevalence of low back pain in the general population. *Spine* 1995; 20, 1889–1894.

Punnett L, Fine LJ, Keyserling WM, Herrin GD, Chaffin DB. Back disorders and nonneutral trunk postures of automobile assembly workers. *Scandinavian Journal of Work, Environment and Health* 1991; 17, 337–346.

Richardson JK, Chung T, Schultz JS, Hurvitz E. A familial predisposition toward lumbar disc injury. *Spine* 1977; 22, 1487–1493.

Rosen NB. Treating the many facets of pain. *Business and Health* 1986 (May); 7–10.

Rossignol M, Suissa S, Abenhaim L. The evolution of compensated occupational spinal injuries: a three year follow-up study. *Spine* 1992; 17, 1043–1047.

Rowe ML. *Backache at Work*. Perinton Press, Fairport, NY, 1983.

Royal College of General Practitioners. *Clinical Guidelines for the Management of Acute Low Back Pain*. London, 1996.

Scholey M, Hair M. Back pain in physiotherapists involved in back care education. *Ergonomics* 1989; 32, 179–190.

Shekelle PG. What role for chiropractic in health care? *New England Journal of Medicine* 1998; 339, 1074–1075.

Snook SH. Approaches to the control of back pain in industry: job design, job placement and education/training. *Spine: State of the Art Reviews* 1987; 2, 45–59.

Snook SH. The control of low back disability: the role of management. In *Manual Material Handling: Understanding and Preventing Back Trauma,* KHE Kroemer, JD McGlothlin, TG Bobick, eds. American Industrial Hygiene Association, 1989.

Snook SH, Campanelli RA, Hart JW. A study of three preventive approaches to low back injury. *Journal of Occupational Medicine* 1978; 20, 478–481.

Snook SH, Webster BS, McGorry RW, Fogleman MT, McCann KB. The reduction of chronic, non-specific low back pain through the control of early morning lumbar flexion: a randomized controlled trial. *Spine* 1998; 23, 2601–2607.

Spitzer WO. Low back pain in the workplace: attainable benefits not attained. *British Journal of Industrial Medicine* 1993; 50, 385–388.

Spitzer WO, LeBlanc FE, Dupuis M. Scientific approach to the assessment and management of activity-related spinal disorders. a monograph for clinicians. report of the Quebec task force on spinal disorders. *Spine* 1987; 12, S5–S59.

U.S. General Accounting Office. *Private Sector Ergonomics Programs Yield Positive Results*, GAO/HEHS-97-163. Washington, DC, 1997.

Videman T, Battie, MC. A critical review of the epidemiology of idiopathic back pain. In *Low Back Pain: A Scientific and Clinical Overview,* JN Weinstein, SL Gordon, SL Gordon (eds). American Academy of Orthopedic Surgeons, Rosemont, IL, 1996.

Von Korff M, Saunders K. The course of back pain in primary care. *Spine* 1996; 21, 2833–2839.

Von Korff M, Deyo R, Cherkin D, Barlow W. Back pain in primary care: outcomes at 1 year. *Spine* 1993; 18, 855–862.

Waddell G. A new clinical model for the treatment of low-back pain. *Spine* 1987; 12, 632–644.

Waddell G. *Epidemiology Review: The Epidemiology and Cost of Back Pain. The Annex to the Clinical Standards Advisory Group's Report on Back Pain.* HMSO, London, 1994.

Waddell G. Low back pain: a twentieth century health care enigma. *Spine* 1996; 21, 2820–2825.

Waddell G. *The Back Pain Revolution*. Churchill Livingstone, Edinburgh, 1998.

Wahlgren DR, Atkinson JH, Epping-Jordon JE, Williams RA, Pruitt SD, Klapow JC, Patterson TL, Grant I, Webster JS, Slater MA. One-year follow-up of first onset low back pain. *Pain* 1997; 73, 213–221.

Webster B. Cost of low back pain. Unpublished data. Weinstein, J., 1998, A 45-year-old man with low back pain and a numb left foot (Clinical Crossroads). *Journal of the American Medical Association* 1996; 280, 730–736.

Weinstein J. Letters: Clinical Crossroads: a 45-year-old man with low back pain and a numb left foot. *Journal of the American Medical Association* 1999; 281, 893–895.

White AA, Gordon SL. Synopsis: workshop on idiopathic low-back pain. *Spine* 1982; 7, 141–149.

Williams DA, Feuerstein M, Durbin D, Pezzullo J. Health care and indemnity costs across the natural history of disability in occupational low back pain. *Spine* 1998; 23, 2329–2336.

Wood DJ. Design and evaluation of a back injury prevention program within a geriatric hospital. *Spine* 1987; 12, 77–82.

Workers' Compensation Monitor. Disability leave: only half of injured workers were contacted by supervisor. *Workers' Compensation Monitor* 1998 (September); 12: 9.

Chapter 12

Psychophysical Studies of Materials Handling

NIOSH Lifting Equation

Stover H. Snook

12.1 INTRODUCTION

One of the primary goals of occupational ergonomics is to reduce musculoskeletal disorders. By definition, ergonomics is designing the job to fit the capabilities and limitations of as many workers as possible. The first step in the ergonomics approach is to evaluate the job; the second step, if necessary, is to redesign the job. If a well-designed job does not reduce musculoskeletal pain, it should certainly reduce the disability. A well-designed job will help the worker cope with the pain, and remain on the job. There is considerable evidence that remaining active and on the job is excellent therapy (Spitzer 1987; Spitzer 1993; Royal College of General Practitioners 1996; Waddell 1996; Nachemson 1983).

12.2 PSYCHOPHYSICAL EXPERIMENTS

A basic question in evaluating a manual handling job is: 'How much should the worker be required to lift?' Psychophysics is a particularly useful tool for answering that question. A series of psychophysical studies at the Liberty Mutual Research Center led to the development of tables for determining the maximum acceptable weights of lift, lower, and carry; and the maximum acceptable forces for pushing and pulling. The independent variables were task height, task distance, object size, frequency, and gender. The dependent variables were object weight, heart rate, and oxygen consumption.

During lifting, lowering, and carrying tasks, subjects handled industrial tote boxes. The boxes varied in length (the distance between the hands) and width (the distance away from the body). When handles were used, they were located mid-way in the width dimension. Each subject varied the weight of the box by adding or subtracting loose lead shot with a small scoop. Welding rod was used in the larger boxes. In an attempt to minimize visual cues, each box contained a false bottom. The subjects were aware of the false bottom, but never knew how much lead shot or welding rod it contained. The amount of weight in the false bottom was randomly varied.

A special device with a rapidly moving shelf was used to automatically lower the box after each lift, or to raise the box after each lowering task. When the box was removed from the shelf by the subject, the shelf quickly moved to a new, predetermined position in time for the subject to place the box back on the shelf.

The box was slid off the shelf, lifted (or lowered) clear of the moving shelf, and then slid back onto the shelf. In most cases, the lifts and lowers were not truly symmetrical since some degree of body twisting was involved. When the box was replaced on the shelf, the shelf returned to its original position. The starting and stopping points of the shelf were adjustable.

Pushing and pulling tasks were simulated on a specially constructed treadmill. The treadmill was powered by the subject as he or she pushed or pulled against a stationary bar. A load cell on the stationary bar measured the horizontal force being exerted. The subject controlled the resistance of the treadmill belt by varying the amount of electric current flowing into a magnet particle brake geared to the rear drum of the treadmill. All subject controls were devoid of visual cues.

All experiments were conducted in a 3.9 m by 2.8 m by 3.0 m environmental chamber. The dry bulb temperature was maintained at a moderate 21.0 degrees C.; relative humidity was 45%.

Subjects were second-shift (evening shift) workers from local industry. They were all given a medical examination prior to their participation in the experiments to ensure that they were in relatively good health. A battery of 41 anthropometric measurements was taken for each subject. Clothing was controlled by providing surgical scrub suits for all subjects. Safety shoes with neolite heels and soles were also provided to control for variations in traction during pushing and pulling tasks. Heart rate was monitored continuously by radio telemetry for each subject. Instantaneous measurements of steady state oxygen consumption were obtained after subjects had selected the maximum acceptable weight or force.

Subjects were instructed to work on an incentive basis, working as hard as they could without straining themselves, or without becoming unusually tired, weakened, overheated, or out of breath. Four or five days of training sessions were provided to allow subjects to gain experience at monitoring their own feelings, and adjusting the object weight or force. Subjects began with moderate frequency, short duration tasks and gradually conditioned themselves to the faster, longer tasks.

New subjects tended to accept the initial weight of the object that was given to them. Therefore, they were encouraged to make adjustments in the weight by starting them with a very light or a very heavy weight. To overcome adaptation effects, each manual handling task was broken up into two 20-minute segments; one segment with a heavy initial weight and the other segment with a light initial weight. There was no rest period between the two segments. If the results of the first segment were within

15% of the second segment, the average of the two results was recorded. Otherwise, the results were discarded and the test re-run at another time.

Test sessions lasted approximately four hours. Each test session usually consisted of five different 40 minute tasks, separated by ten minute breaks. Groups of three subjects participated in at least two test sessions per week for a minimum of ten weeks.

12.3 PSYCHOPHYSICAL TABLES

The results from the individual experiments were published in the technical literature. However, in order to facilitate the application of results to actual manual handling tasks, the results were integrated into a single set of tables in 1978 (Snook). After an additional twelve years of testing, a revised set of tables was published in 1991 (Snook and Ciriello). The 1991 publication contains the following nine tables for evaluating manual handling tasks:

- Table 2: Maximum acceptable weight of lift for males.
- Table 3: Maximum acceptable weight of lift for females.
- Table 4: Maximum acceptable weight of lower for males.
- Table 5: Maximum acceptable weight of lower for females.
- Table 6: Maximum acceptable forces of push for males.
- Table 7: Maximum acceptable forces of push for females.
- Table 8: Maximum acceptable forces of pull for males.
- Table 9: Maximum acceptable forces of pull for females.
- Table 10: Maximum acceptable weight of carry (male and female).

Three steps and seven items of information are needed to use the tables.

Three steps	Seven items of information
Identify the correct table	Type of task Gender of worker
Identify the correct 5 by 8 matrix (frequency × population percent)	Height of task Width of object Distance of movement
Identify the correct value	Frequency Population percent

The width of the object is necessary only for lifting and lowering tasks.

It is important to note that Tables 2, 3, 4, 5, and 10 are based on handling boxes with handles, and handling boxes close to the body. The values in the tables should be reduced by approximately 15% when handling boxes without handles, and by approximately 50% when handling smaller boxes with extended horizontal reaching between knee height and shoulder height.

It is also important to note that some of the weights and forces in Tables 2–10 will exceed recommended physiological criteria when performed continuously for eight hours or more. The recommended eight-hour criteria are approximately 1000 ml/min of oxygen consumption for males; 700 ml/min, for females. (NIOSH 1981). Weights and forces that exceed these criteria are italicized in Tables 2–10.

Tables 2–10 give maximum acceptable weights and forces for individual manual handling tasks or components (e.g. lifting). Frequently, however, industrial tasks involve combinations with more than one component (e.g. lifting, carrying, and lowering). Each component of a combined tasks should be analysed separately using the frequency of the combined task. The weight or force of the component with the lowest percent of population represents the maximum acceptable weight or force for the combined task. However, since the physiological cost of the combined task will be greater than the cost for individual components, it should be recognized that some of the combined tasks may exceed recommended physiological criteria for extended periods of time (NIOSH 1981).

12.4 THE NIOSH LIFTING EQUATION

The rationale and criterion for developing the NIOSH lifting equation is found in Waters et al. 1993. The equation is based upon three criteria: biomechanical, physiological, and psychophysical. The specific criteria are shown in Table 12.1.

Table 12.1 Criteria for the NIOSH Lifting Equation.

Discipline	*Design criterion*	*Cut-off value*
Biomechanical	Maximum disc compression force	3.4 kN
Physiological	Maximum energy expenditure	2.2–4.7 kcal/min*
Psychophysical	Maximum acceptable weight	Acceptable to 75% of females and 99% of males

* Depending on the vertical height and duration of lift.

The NIOSH lifting equation (12.1) determines the Recommended Weight Limit (RWL). The RWL is the weight that nearly all healthy workers can perform over a substantial period of time (up to 8 hours) without an increased risk of developing lifting-related low back pain. The RWL is defined by the following equation:

$$RWL = LC \times HM \times VM \times DM \times AM \times FM \times CM \quad (12.1)$$

where: LC = Load constant = 23 kg
HM = Horizontal multiplier = 25/H
VM = Vertical multiplier = $1-(0.003|V-75|)$
DM = Distance multiplier = $0.82+(4.5/D)$
AM = Asymmetric multiplier = $1-(0.0032A)$
FM = Frequency multiplier (from Table 16.2)
CM = Coupling multiplier (from Table 16.3)

where: H is the distance of the hands away from the mid-point between the ankles (cm).
V is the distance of the hands from the floor (cm).
D is the vertical travel distance between the origin and the destination of the lift (cm).
A is the angular displacement of the load from the sagittal plane (degrees).
F is the average frequency rate of lifting measured in lifts/min.

The load constant is derived from the maximum acceptable weight of lift for 75% of female workers (and about 90% of male workers) under optimal conditions (i.e. occasional lifting, small object, short distance, good handles, etc.).

The NIOSH lifting equation assumes that lifting and lowering tasks have the same level of risk for low back disorders. The equation does not apply if any of the following occur:

- Lifting/lowering with one hand.
- Lifting/lowering for over eight hours.
- Lifting/lowering while seated or kneeling.
- Lifting/lowering in a restricted work space.
- Lifting/lowering unstable objects.
- Lifting/lowering while carrying, pushing, or pulling.
- Lifting/lowering with wheelbarrows or shovels.
- Lifting/lowering with high speed motion (faster than 30 per second).
- Lifting/lowering with unreasonable foot/floor coupling (<0.4 coefficient of friction).
- Lifting/lowering in an unfavourable environment (temperature outside 19–26 degrees C; humidity outside 35–50%).

The Lifting Index (LI) (12.2) is the weight of the load lifted (L) divided by the Recommended Weight Limit (RWL):

$$\text{Lifting index (LI)} = \frac{\text{Load weight (L)}}{\text{Recommended weight limit (RWL)}} = \frac{L}{RWL} \quad (12.2)$$

Table 12.2 Frequency multiplier.

	Work duration					
	<1 Hour		*<2 Hours*		*<8 Hours*	
Frequency (lifts/minute)	*V< 75*	*V >75*	*V<75*	*V >75*	*V <75*	*V>75*
0.2	1.00	1.00	0.95	0.95	0.85	0.85
0.5	0.97	0.97	0.92	0.92	0.81	0.81
1	0.94	0.94	0.88	0.88	0.75	0.75
2	0.91	0.91	0.84	0.84	0.65	0.65
3	0.88	0.88	0.79	0.79	0.55	0.55
4	0.84	0.84	0.72	0.72	0.45	0.45
5	0.80	0.80	0.60	0.60	0.35	0.35
6	0.75	0.75	0.50	0.50	0.27	0.27
7	0.70	0.70	0.42	0.42	0.22	0.22
8	0.60	0.60	0.35	0.35	0.18	0.18
9	0.52	0.52	0.30	0.30	0.00	0.15
10	0.45	0.45	0.26	0.26	0.00	0.13
11	0.41	0.41	0.00	0.23	0.00	0.00
12	0.37	0.37	0.00	0.21	0.00	0.00
13	0.00	0.34	0.00	0.00	0.00	0.00
14	0.00	0.31	0.00	0.00	0.00	0.00
15	0.00	0.28	0.00	0.00	0.00	0.00
>15	0.00	0.00	0.00	0.00	0.00	0.00

Table 12.3 Coupling multiplier.

	V<75	*V >75*
Couplings	*Coupling multipliers*	
Good	1.00	1.00
Fair	0.95	1.00
Poor	0.90	0.90

The LI can be used to estimate the relative magnitude of physical stress for a lifting task. The greater the LI, the smaller is the percentage of workers that can safely perform the task. The LI is a good way to compare two or more task designs.

For further information, consult the NIOSH Applications Manual (NIOSH 1994). Procedures for analysing multiple or combination tasks are also found in the Applications Manual.

12.5 VALIDATION

A recent study by Marras et al. (1999) investigated the effectiveness of the 1981 NIOSH Work Practices Guide for Manual Lifting, the 1993 NIOSH

Lifting Equation, and the psychophysical tables in correctly identifying jobs with high, medium, and low risk of low back disorders. The study used a database of 353 industrial jobs representing over 21 million person-hours of exposure. The results indicated that all three methods were predictive of low back disorders, but in different ways. Table 12.4 depicts the percentage that each method correctly predicted high, medium, and low risk jobs.

Table 12.4 Correct identification of jobs with high, medium, and low risk of low back disorders by three different assessment methods (from Marras et al. 1999).

	NIOSH 81	*NIOSH 93*	*Psychophysical*
High-risk jobs (sensitivity)	10%	73%	40%
Medium-risk jobs	43%	21%	36%
Low-risk jobs (specificity)	91%	55%	91%

The 1981 NIOSH Work Practices Guide underestimated the risk by predicting that most jobs were low risk (low sensitivity, high specificity). The 1993 NIOSH Lifting Equation overestimated the risk by predicting that most jobs will be high risk (high sensitivity, low specificity). The psychophysical tables fell somewhere in between.

12.6 REFERENCES

Marras WS, Fine LJ, Ferguson SA, Waters TR. The effectiveness of commonly used lifting assessment methods to identify industrial jobs associated with elevated risk of low-back disorders, *Ergonomics* 1999; 42, 229–245.

Nachemson A. Work for all. *Clinical Orthopaedics and Related Research* 1983; 179, 77–85.

National Institute for Occupational Safety and Health (NIOSH). *Work Practices Guide for Manual Lifting,* DHHS (NIOSH) Publication No. 81-122. NIOSH, Cincinnati, OH, 1981.

National Institute for Occupational Safety and Health (NIOSH). *Applications Manual for the Revised NIOSH Lifting Equation,* DHHS (NIOSH) Publication No. 94-110. NIOSH, Cincinnati, OH, 1994.

Royal College of General Practitioners. *Clinical Guidelines for the Management of Acute Low Back Pain.* London, 1996.

Snook SH. The design of manual handling tasks. *Ergonomics* 1978; 21, 963–985.

Snook SH, Ciriello VM. The design of manual handling tasks: revised tables of maximum acceptable weights and forces. *Ergonomics* 1991; 34, 1197–1213.

Spitzer WO. Low back pain in the workplace: attainable benefits not attained. *British Journal of Industrial Medicine* 1993; 50, 385–388.

Spitzer WO, LeBlanc FE, Dupuis M et al. Scientific approach to the assessment and management of activity-related spinal disorders: a monograph for clinicians: report of the Quebec task force on spinal disorders. *Spine* 1987; 12, S5–S59.

Waddell G. Low back pain: a twentieth century health care enigma. *Spine* 1996; 21, 2820–2825.

Waters TR, Putz-Anderson V, Garg A, Fine LJ. Revised NIOSH equation for the design and evaluation of manual lifting tasks. *Ergonomics* 1993; 36, 749–776.

Chapter 13

Biomechanical Models in High Exertion Manual Jobs

Don B. Chaffin

13.1.1 INTRODUCTION

Biomechanical models of human exertions have been developed by many different researchers over the last five decades as reviewed in Chaffin et al. (1999). These have become quite sophisticated, and when combined with human performance and population tissue failure data, the models become valid tools to understand and predict the risk associated with specific types of exertions. In this context they have construct validity, in that they are based on well recognized concepts and data in both mechanics and biology. In addition, prospective longitudinal studies, to be presented, have indicated that when high exertion task requirements are compared with model predicted population strength capabilities, those jobs with high peak exertion requirements are associated with significantly elevated musculoskeletal injury rates, thus establishing their empirical validity.

13.1.2 WHY USE HUMAN SIMULATION TECHNOLOGY

Manual exertions remain an essential part of most jobs today, despite automation in many industries. In 1981 it was estimated that one-third of jobs in the United States contained manual exertions with significant strength requirements (NIOSH 1981). Though many jobs have been redesigned to reduce such exertions in hard goods manufacturing, there is no evidence that such is the case in the service sector. These types of exertions are believed by many to account for a disproportionate number of serious injuries, mostly to the musculoskeletal system.

One of the major requirements in preventing work related musculoskeletal injuries is to know precisely what strenuous exertion requirements are contained in a job. This requires improved job analysis tools which can accurately predict potentially harmful stresses to the musculoskeletal system. But to simply depict the level of stress at a joint, or in a muscle or tendon is not enough to motivate job changes. The stress levels must be evaluated, i.e. compared to a population's performance capacity or injury risk.

In this context, the development of computerized biomechanical models of the human musculoskeletal system are important to ergonomics. These software programs are not just research tools for the laboratory, or to depict unusual looking human forms and movements for television commercials and motion pictures. Some of these models provide a means to simulate a variety of exertions and predict the consequences for a specified population. This is a goal that is now well recognized in the design of vehicle interiors, and a variety of human simulation tools for this purpose have been developed (Peacock and Karwowski 1993).

The urgency to develop such comprehensive job evaluation methods is not only based on the need to contain the costs and suffering of musculoskeletal injuries, but also to understand how better to design workplaces to accommodate older workers, women, and physically impaired job applicants and injured workers. Unfortunately, from inspection of various workplaces it has been reported that human manual exertion requirements are not often considered *early* in the design process (Evans and Chaffin 1985). To improve the design of manual tasks, easy to use manual exertion simulation CAD tools are needed. One of the biggest challenges in this regard will be to predict accurately a variety of human postural and motion requirements in different work settings, and for a variety of populations. What follows is a general description of this problem and the software developments at the University of Michigan's Center for Ergonomics to address this challenge. It is hoped that our experiences with this general problem will serve as a valuable reference to others who believe in the philosophy that human simulations can improve the use of ergonomics in workplace design.

13.1.3 BACKGROUND

One of the major issues confronting a job analyst or job designer is whether the manual task evaluation should be concerned with a particular exertion in a job or with a sequence of exertions. The input and output may be quite different, as depicted in Table 13.1.

As inspection of Table 13.1 reveals, analysis of a single static exertion requires the minimum amount of information about the task, but is still quite demanding of a job analyst/designer. In particular, the postural requirements must be accurately portrayed, as well as both the hand force (and particularly the direction of the hand force in each hand) and anthropometry of the person being simulated. If the posture and hand forces are depicted well, static biomechanical models exist which can reasonably predict population strength, reach, balance and foot slip potential (Chaffin et al. 1999).

The addition of human motion (i.e. dynamics) into the job analysis is problematic at this time however, in that motions can induce peak inertial

Table 13.1 Single v. multiple exertion task input/output requirements.

Type of exertions	*Typical input data*	*Typical output provided*
Single static exertion	1 Posture required (hand location) 2 Hand force (mag. and direction) 3 Anthropometry (population)	% Population strength capability Low-back force tolerance Body balance and reach capability Static foot COF
Single dynamic exertion	1–3 Same as static 4 Body segment motion parameters	Motion kinematics and dynamic strengths Body balance Dynamic foot COF Motion time estimates
Multiple dynamic exertions	1–4 Same as single dynamic 5 Temporal sequence of motions 6 Work/Rest schedule	Job metabolic energy required Job performance time prediction Repetitive motion injury rating

loads on the body, as well as alterations in assumed body postures, which will, on average, cause the static analyses to error by ±30% or more. The difficulty in using dynamic analyses is two-fold: First, there does not exist a method of accurately predicting (as opposed to measuring) the average postural and inertial loading effects for different tasks without expensive equipment and time consuming analyses. It has only been in the last few years that a systematic attempt to model and predict normal human motions has begun (Chaffin, Faraway and Zhang 1999). Second, dynamic strength and tissue tolerance data for the working population do not exist, and hence only qualitative recommendations are possible for dynamic exertions. This is not to imply that dynamic analyses are not valuable at this time, especially for evaluation of existing work situations. Indeed, with the advent of automated video image analysis, body segment motion data can be acquired by direct measurement of a person performing a job. These data then can be evaluated by dynamic biomechanical models with the resulting muscle and joint force predictions being useful in guiding the improvement of a job, through comparative evaluations of moments and forces at different joints.

Traditionally the interest in multiple task analysis has been to either predict the time required to perform the job, or the metabolic energy required by an average person performing the job. More recently, sequential task and postural data have been combined empirically to predict the risk of a variety of chronic musculoskeletal complaints and injuries believed to be related to excessive repetition of forceful exertions in certain postures (Keyserling 1986). If the job analyst has the time and ability to observe and record a worker performing a job in question, then all three types of analyses described earlier in Table 13.1 are feasible (i.e. specialized software is available to assist in developing the outputs described). If, however, the analyst does not have the time and expertise to develop the input data, or the job isn't accessible (or doesn't even exist) then the analyst must rely on the fidelity and validity of computerized manual task simulations to develop useful design recommendations. Figure 13.1 describes the software developed (and being developed) at the University of Michigan for these various applications.

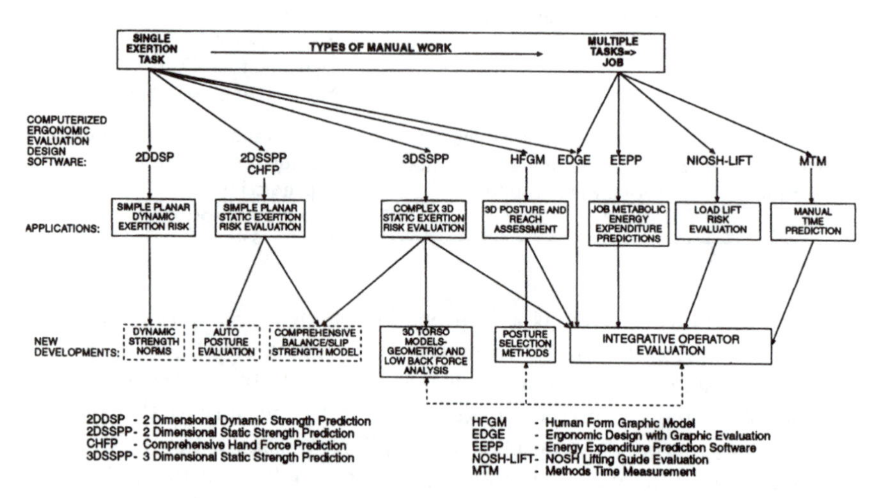

Figure 13.1 An overview of UM software for manual work evaluation and design.

Because the software available for a single exertion static analysis is well developed, and is used for performing both job evaluations and new job design analysis, it will be further described to illustrate the challenges and opportunities provided by combining human form graphic modelling with biomechanical modelling. In particular, what follows is a brief description of the 3D Static Strength Prediction Program development, and how it is being improved for design simulations by the incorporation of a new Human Form Graphic Model and posture prediction model developed recently by Beck and Chaffin (1992).

The reasons it was deemed necessary to include these latter improvements are:

1. To simulate a 3D static exertion with the present software may require too much time in data entry tasks (11 minutes), thus discouraging widespread use by an analyst.
2. An excessive long time to input data discourages a designer of a new job from being creative (trying many alternatives), and in fact, may induce

excessive errors in the postures finally used in the biomechanical simulations.

3. The job designer or analyst may have little experience or knowledge regarding human postures, and thus he or she will require computer assistance in choosing postures for specific job design requirements.

13.2 OVERVIEW OF BIOMECHANICAL STRENGTH PREDICTION MODEL

The strength prediction model denoted as 2DSSPP® and 3DSSPP® in Figure 13.1 considers the following four different types of limitations in predicting how much force a population strata can be expected to exert in a particular hand position:

1. Population isometric muscle strength data at each major skeletal joint.
2. Low back (L5/S1 and L4/L5) motion segment compression force tolerances.
3. Foot slip static coefficient-of-friction (COF) limits.
4. Body balance capability.

What follows is a brief explanation of how the first two physical limitations are computed in these programs. A more thorough review is presented in Chaffin et al. (1999).

Muscle strength limits

To compute these limits the following logic is used, and is briefly described (Figure 13.2):

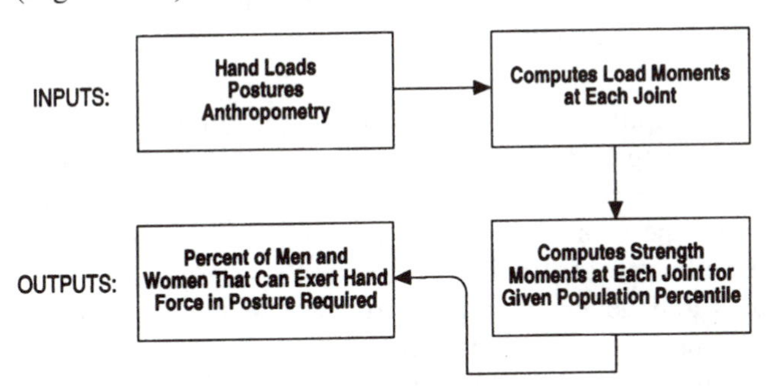

Figure 13.2 Logic for muscle strength prediction in 2D and 3DSSPP®.

To compute joint moments simple Newtonian Mechanics are used, wherein forces (i.e. body segment weights and hand loads) are multiplied by their perpendicular distance from joint centres (i.e. moment arms). The following (Figure 13.3) is an example computing elbow load moment ME while holding a weight.

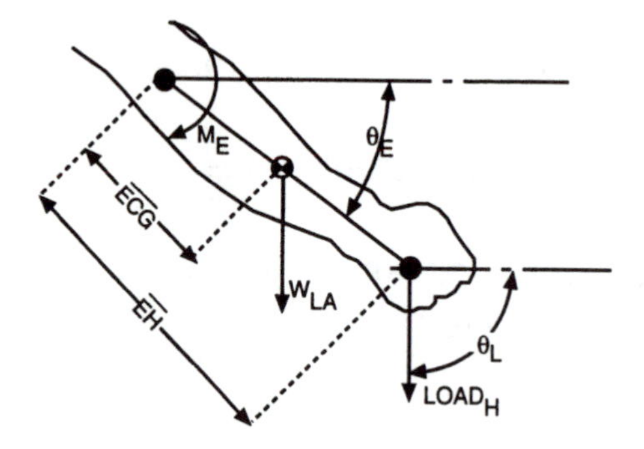

where $M_E = \cos\theta_E[(\overline{ECG} \times W_{LA}) + (\overline{EH} \times LOAD_H)]$

θ_E	is angle at elbow measured from horizontal to segment centreline
$\overline{ECG}$	is distance from elbow to lower arm and hand CG
W_{LA}	is weight of lower arm and hand
$\overline{EH}$	is distance from elbow to hand grip centre
$LOAD_H$	is the force at the hand grip centre
θ_L	is angle of hand force from horizontal

Figure 13.3 Elbow moment estimating procedure (from Chaffin et al. 1999).

Note that only three types of input data are required:

1. Hand forces ($LOAD_H$)
2. Anthropometry, which provides segment dimensions and weights.
3. Postures, which provide joint angle data.

A recursive method of computation is used to compute the moments at all of the major joints in the linkage structure shown in Figure 13.4. The resulting joint moments are compared with static strength data derived from isometric strength tests performed on approximately 3000 workers and college students (Chaffin et al. 1987). These strength test data were regressed on postures and gender strata to provide statistical prediction equations for each muscle function at the major joints. The resulting values are then systematically compared with the moments predicted at each joint when a person is simulated performing a specific exertion (i.e. lift, push, pull, etc.) with one or both hands.

The result is a tabulation of the percentage of the population capable of performing a specific exertion for each joint and muscle function as shown in Figure 13.5. For example, in Task 1 with a low moment load at the joint

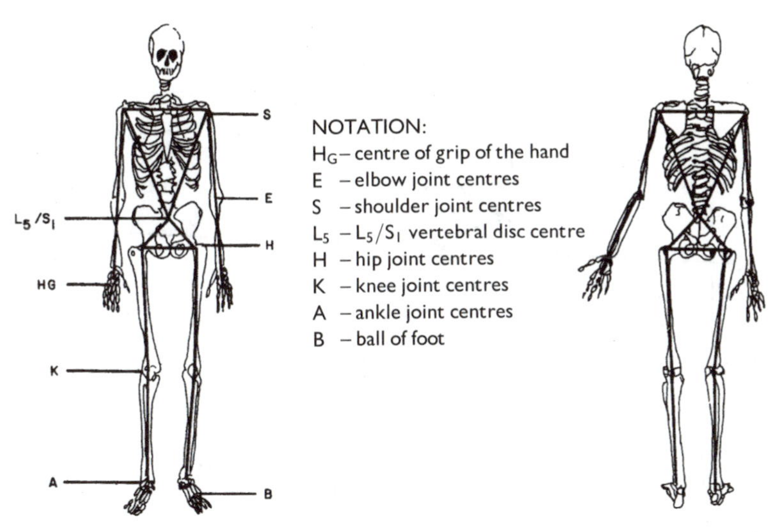

Figure 13.4 Linkage and joint centre representation used in the UM's biomechanical 3D Static Strength Prediction Program(TM).

being analysed, the exertion can be performed by all men and over 90% of women. But for Task 2, with a higher moment loading, the exertion can only be performed by about 65% of men and only 15% of women.

Predictions of this type have been validated by comparing the model outputs with the strength of people performing a variety of lifting,

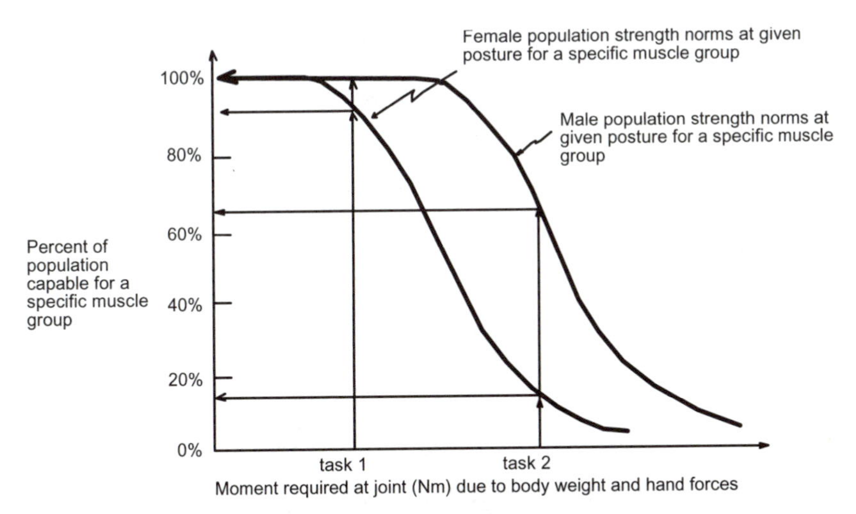

Figure 13.5 Depiction of percentage of male and female populations predicted to have the specific muscle strength capability needed to perform two tasks having different joint moment loadings.

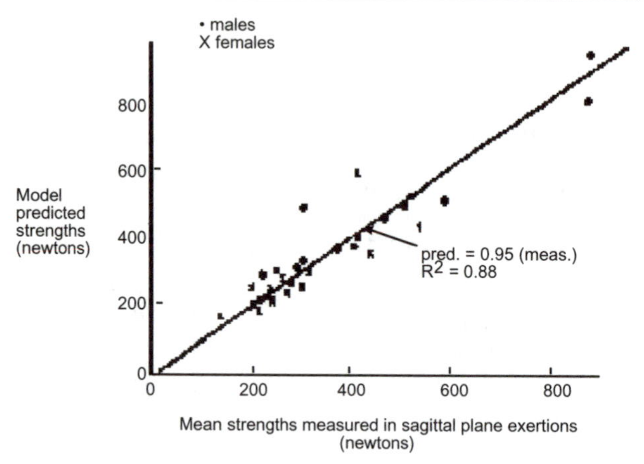

Figure 13.6 Comparison of 50 percentile strength values with model predicted strengths using 50 percentile muscle strengths at each joint, as summarized in Chaffin et al. (1999).

pushing and pulling tasks (Chaffin et al. 1987). Correlations show best results for 2D symmetric tasks ($r^2 > 0.85$) in Figure 13.6.

The model of job predicted strength requirements was also used in the 1970s as the basis for two separate prospective, incumbent, epidemiological studies in several different plants. In these studies jobs were first evaluated by using the strength model predictions to determine the peak exertion requirements in each job. Once identified, these high exertion tasks were replicated in the medical department using a specially built isometric strength tester. Then 1051 workers employed on the jobs were strength tested and their medical status was monitored for the next one to two years. The results are shown in Figures 13.7 and 13.8.

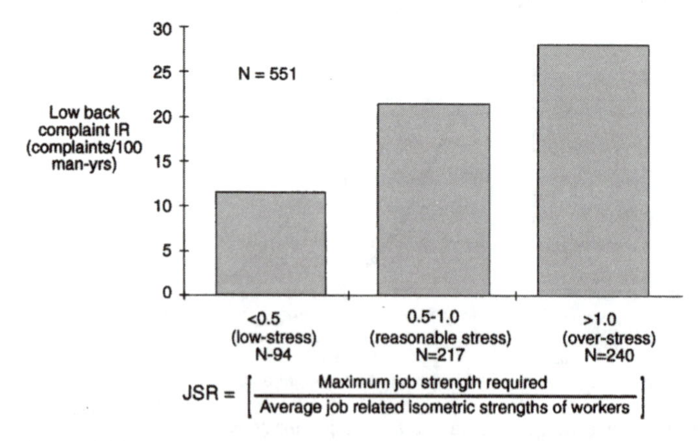

Figure 13.7 Prospective Incumbent Study– Low back pain complaints of incumbent workers with different job related strengths (1–2 year follow-up) (Chaffin et al. 1978).

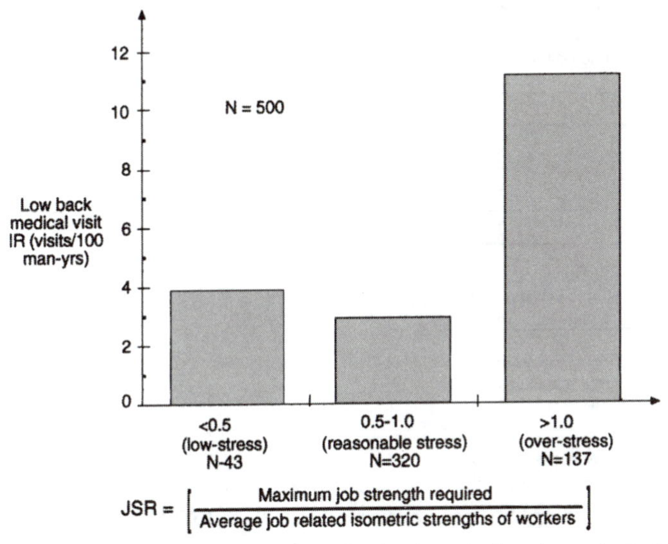

Figure 13.8 Prospective Incumbent Study– Low back pain medical visits for incumbent workers with different job related strengths (1–1/2 year follow-up) (Chaffin et al. 1974).

Those workers who tested weaker than the peak job requirements (wherein the ratio of job strength requirement JSR was over 1.0) had about 2.5 times as many low back complaints and medical visits as their stronger peers). The conclusion being that if the job peak exertion requirements exceed the strength capabilities of workers (as determined by job specific strength tests) the incidence rate of both complaints and medical visits for low back pain will significantly increase.

Low back compression force limit

The logic for computing the L5/S1 motion segment compression force within the 3DSSPP software is shown in Figure 13.9. The logic assumes that once the L5/S1 moment is computed (as described in the preceding), the torso muscles contract to stabilize the column. In the 2DSSPP logic for sagittal plane exertions a single torso muscle equivalent contraction force is included as shown in Figure 13.10.

When added to body segment weights and hand forces (with a minor adjustment for abdominal pressure effects) the above logic provides the basis for predicting the compression force on the L5/S1 disc. In the 3DSSPP several different torso muscle contraction forces are allowed. These are depicted in Figure 13.11. The forces are partitioned using optimization procedures to minimize the intensity of muscle contractions and resulting spinal compression forces. The values so computed are

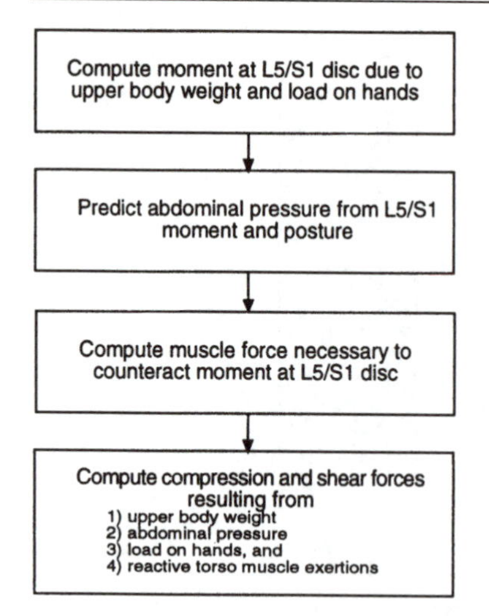

Figure 13.9 Logic for computing L5/S1 compression forces in 2D and 3DSSPP®.

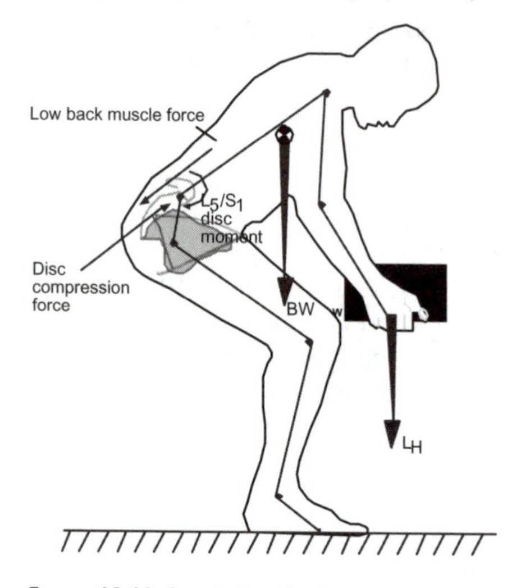

Figure 13.10 Simple low-back model of lifting for static coplanar lifting analyses. The load on the hands L_H and torso and arm weights BW act to create moments at the L5/S1 disc of the spine. The moments are resisted primarily by the back muscles. The high muscle forces required in such a task cause high disc compression forces.

compared to published cadaver motion segment failure levels. In this context the NIOSH spinal risk of 3400N is often used as a population norm to protect most workers.

Because the low back muscles act close to the motion segment rotation axis (near centre of disc), they create a high compression force on the disc (Figure 13.12).

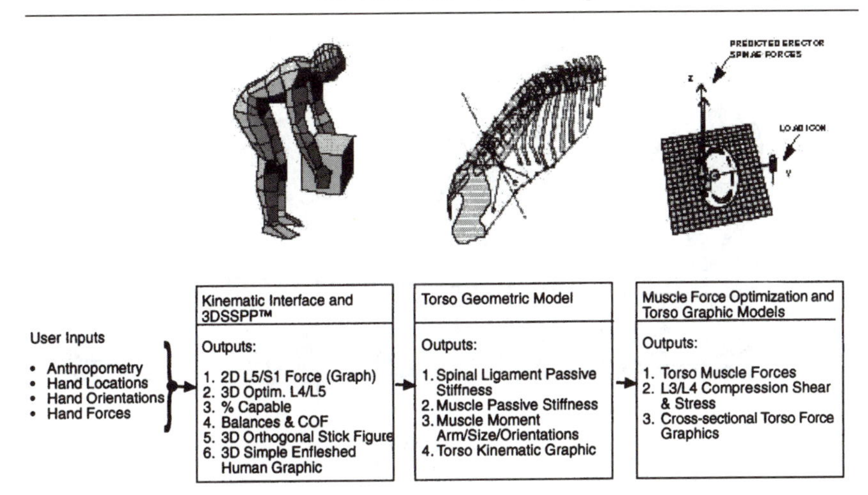

Figure 13.11 Software developed to allow a person to enter 3D data describing an exertion (on left) and receive tabular and graphical information about predicted muscle and spinal forces (on right).

As expected, based on the preceding biomechanical principles, the horizontal distance to the load H is just as important as the magnitude of the load held in the hands. Thus, to minimize both soft tissue stresses and disc compression forces, loads need to be kept close, and the torso kept as vertical as possible. The H distance factor is even more important when the load is closer to the floor. If the load has to be moved around the knees (because of its size) it is better to move closer to it, and use a stoop-lift, rather than squat down and lift it. The latter type of lift can create a large H effect which increases the compression forces. This is depicted when lifting a 90-pound, large size (wide) object. What also is depicted in the example shown in Figure 13.13 is that only about 31% of

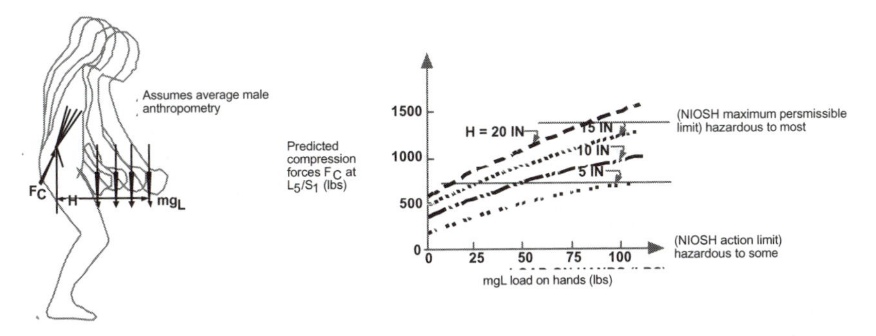

Figure 13.12 Predicted L5/S1 disc compression forces for varying loads lifted in four different positions from body.

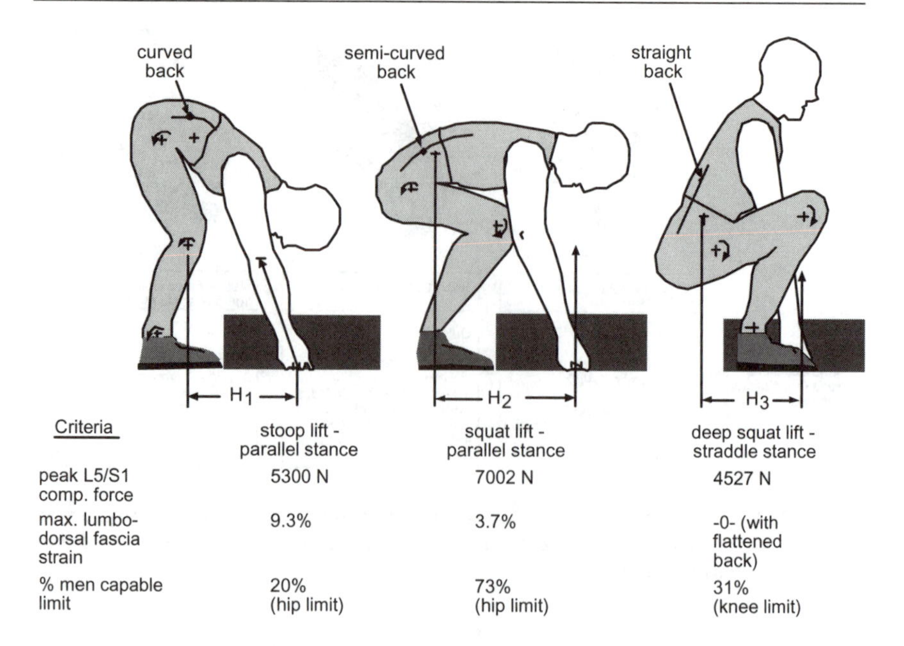

Criteria	stoop lift - parallel stance	squat lift - parallel stance	deep squat lift - straddle stance
peak L5/S1 comp. force	5300 N	7002 N	4527 N
max. lumbo-dorsal fascia strain	9.3%	3.7%	-0- (with flattened back)
% men capable limit	20% (hip limit)	73% (hip limit)	31% (knee limit)

Figure 13.13 Initial lifting postures are important lifting 400 N (90 lb) bulky box (24″ × 26″ D) (Anderson and Chaffin 1986).

men would have sufficient knee extension muscle strength to lift the load using a squat lift. Therefore, though it is better on the back to straddle the load and lift with your legs, many people will not have sufficient leg strength to do so.

Excessive disc compression forces are believed to cause increased risk of disc degeneration and ultimate failure, as shown in Figure 13.14.

A number of different studies have been conducted on cadaver spines to determine the maximum lumbar motion segment compression force limits. A composite of the results is shown (Figure 13.15). Clearly, the values from the studies are not all the same. NIOSH risk levels also are indicated on the graph.

One factor that decreases spinal compression force limits is age. As depicted by NIOSH (1981), on average, lumbar motion segments over 50 years old are about 45% weaker than segments under 40 years, for both men and women. The NIOSH Action Limit defined in 1981 for compression failure is 3400 N (3.4 KN), which is about 760 lb. As was shown earlier in Figure 13.12, lifting a 200 N (44 lb) near knuckle height can easily create compression forces of this magnitude.

The need for concern over such spinal force loadings is confirmed by increased low-back pain medical incidents associated with such lifting, as shown in Figure 13.16. Lifting loads that exceed the NIOSH 3.4 KN (760 lb)

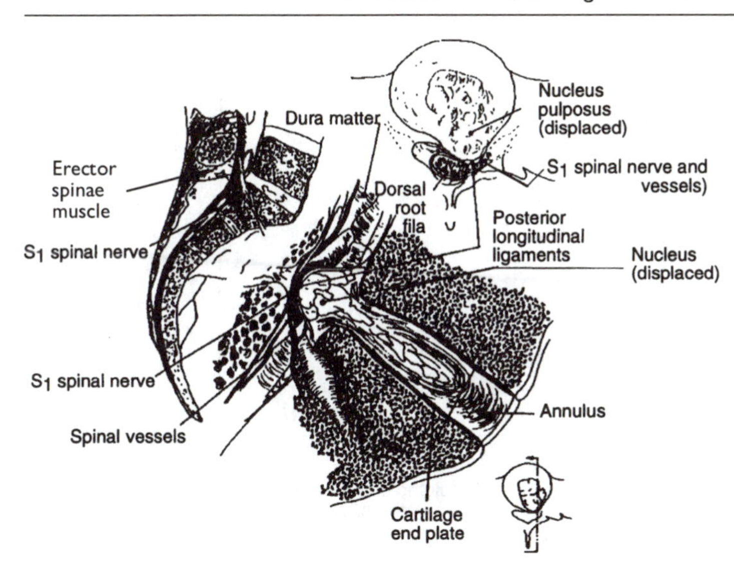

Figure 13.14 Example of L5/S1 disc herniation resulting in spinal nerve root compression and inflammation.

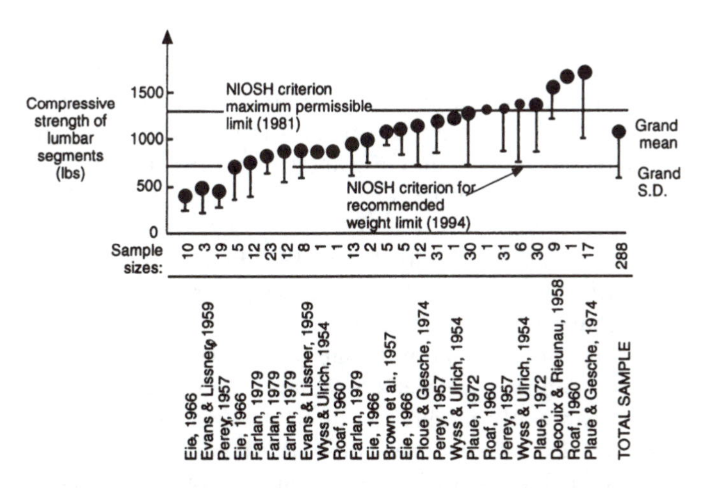

Figure 13.15 Composite of cadaver studies to determine the compression strengths of lumbar segments (adapted from Jäger 1987).

limit appear to create about a 3× greater risk than exists in less stressful tasks. Also, if the compression force requirements of a job are greater than the NIOSH (1981) 6.4 KN (1430 lb) Maximum Permissible Limit for the spine, there appears to be an 8× greater relative risk.

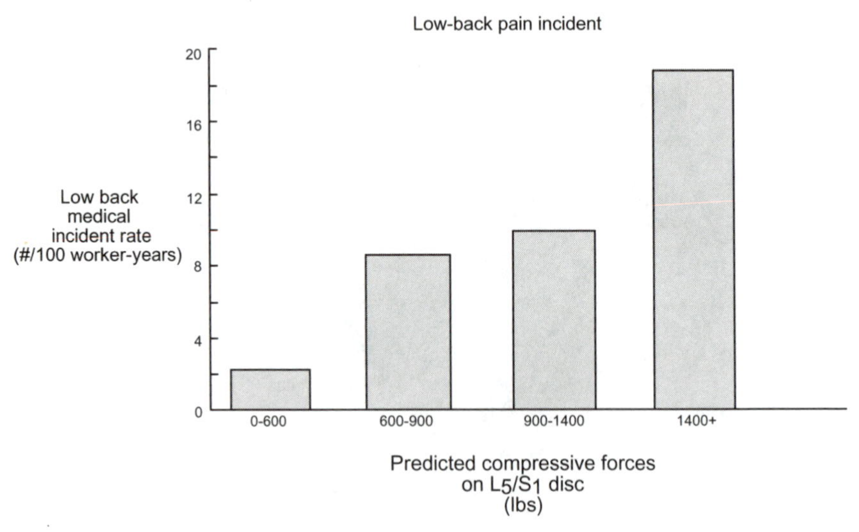

Figure 13.16 Relation between LBP and Peak Compressive Forces on Jobs (NIOSH 1981, based on data from Chaffin and Park 1973).

13.3 CURRENT STATIC STRENGTH PREDICTION PROGRAM®

The preceding physical limits for static exertions were implemented beginning in 1969 on main frame computers. The first programs were two dimensional, but by 1975 a 3D version for muscle strength prediction was operating at the University of Michigan (Garg and Chaffin 1975), with a 3D low-back torso model being added in 1988 based on the work of Bean et al. 1988.

The 2DSSPP® software has been licensed since 1983 by the University to run on an IBM-PC platform. The main screen for this program is shown in Figure 13.17. The emphasis on this version of the model was to provide a simple to use and very fast job evaluation design tool. Since only five body angles are required to define a worker's posture, along with a single handforce, the analyst can easily modify the input data on the upper left screen, and visualize the effects via changes in the stick figure configuration, percent capable graphs, back compression force graphs, foot coefficient-of-friction values, and body balance warnings. Though many iterations of joint angles and hand forces may be necessary to solve a particular exertion problem, the computational speed (less than 1 s on an older 286 machine) has allowed many different users to benefit from this simple tool.

The 3D Static Strength Prediction Program has been licensed by the University of Michigan since 1989. The main input and output screens are depicted in Figure 13.18. The program is best implemented on at least an IBM-386 platform with math co-processor and as a Windows application.

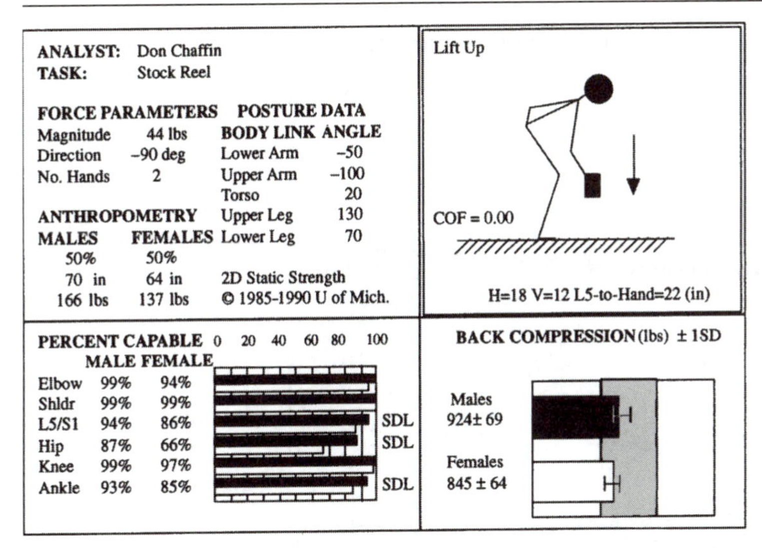

Figure 13.17 Main screen for UM's 2D Static Strength Prediction Program^{TM}.

Like the 2DSSPP^{TM}, postures and individual hand forces were inputted in earlier DOS releases from the keyboard or from a menu of preset common postures.

One of the most difficult tasks for a user when performing a 3D analysis is to input the correct posture. The use of three orthogonal views of a stick figure requires an analyst to perform significant mental image rotation to rectify the joint angle input data to a non-orthogonal photograph of a worker. Such rotation can add to joint angle errors as discussed by Liu et al. (1997). This level of error can cause the strength predictions to be in error by ±30% or more (Chaffin and Erig 1991). To reduce this type of input error, four features have been recently added to improve this program:

1. Human form graphics, which will allow the analyst to quickly visualize complex, non-orthogonal postures (including a primitive CAD routine for drawing objects around the person).
2. A postural prediction and manipulation method by which the analyst can obtain from the computer program a first approximation of a feasible (in-balance) posture for a person when inputting only hand locations and forces, as well as a person's simple anthropometry.
3. An inverse kinematic method which allows the changing of one joint angle without relocating the hand positions, thus allowing mouse input, point and click modifications to the computer generated postures for a specific workplace task.
4. A means to easily view the computer generated hominoid from the same angle and perspective as a video of a job, thus allowing better postural alignment.

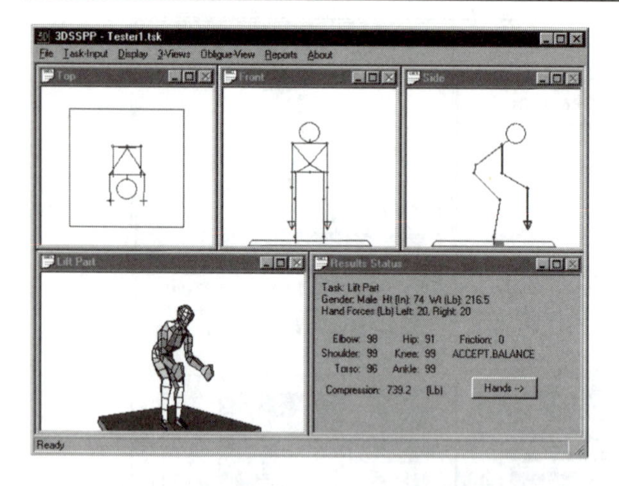

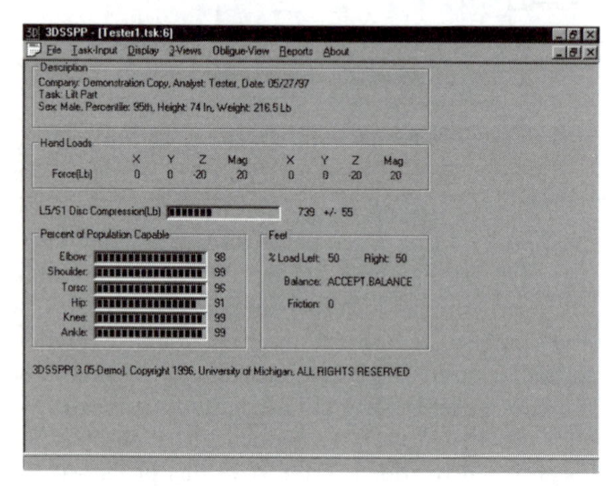

Figure 13.18 Input (top) and output (bottom) screens from 3DSSPP®.

To develop these features the following events were necessary. Beck (1992) developed both a quick draw, human form (enfleshed) graphic model to improve the visualization of non-orthogonal postures, and a behavioural based inverse kinematic posture prediction scheme described below. Based on this work, a new version (V2.0) of 3DSSPP® was developed. Its graphic humanoid is shown in Figure 13.19.

This graphic model was programmed to be easily rotated and zoomed, and will depict a 3D posture in less than 1 s on a contemporary PC with at least a Pentium MMX processor. The behavioural based inverse kinematic method of predicting initial postures is based on previous work by Kilpatrick (1970), Snyder et al. (1972), Park (1973) and Byun (1991) wherein different subjects were photographed while assuming a large

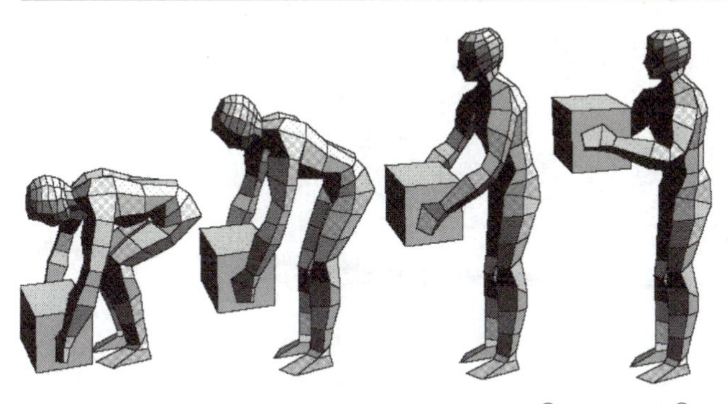

Figure 13.19 UM 3D Human Form Graphic Model™ in 3DSSPP™ V2.0 shown lifting a load.

variety of postures. The digitized photographs were used to generate posture prediction equations which relate Cartesian coordinates of body reference points to hand locations while the subjects were in their 'preferred postures'. Beck used these equations as the basis for choosing body linkage configurations required to set the hands in a specific location. By using an inverse kinematic structure, as a particular segment is moved in space the other segments in the kinematic chain move to the degree necessary to keep the hands in their original location. Imposing this structure not only allowed the computer program to find feasible and preferred postures, but also allows the analyst/designer to point to a segment and move it to another location without losing a desired hand location.

Beck (1992) was able to show that by using both the improved Human Form Graphic Model and the mouse driven, behavioural inverse kinematic posture prediction scheme, novice job analysts were able to reduce their postural entry time by 57%, and were 11% more accurate than when using the previous keyboard data entry scheme in version 1.0 of 3DSSPP™.

Because workplace design is often accomplished within a CAD framework today, a recent innovation was to integrate the 3DSSPP™ software within one widely used PC-CAD application known at AutoCAD™. This allows geometric data about objects and the workplace layout to be easily accessed to position the 3DSSPP hominoid. An example of such an application was recently reported by Feyen et al. (1999). Figure 13.20 depicts such an application for a person lifting transmission torque converters from storage racks with a mechanical balance arm.

Another application of 3DSSPP™ was to determine the effects of lifting a large stock reel weighing 196 N to a spindle about shoulder high. Both low-back compression forces and percent of population capable were predicted for both an average anthropometric male and female. A variety of postures were simulated. The results are graphed in Figure 13.21, and disclose that the low-back compression forces would exceed the NIOSH recommended threshold of 3400 N at both the beginning and ending of the lift, while the

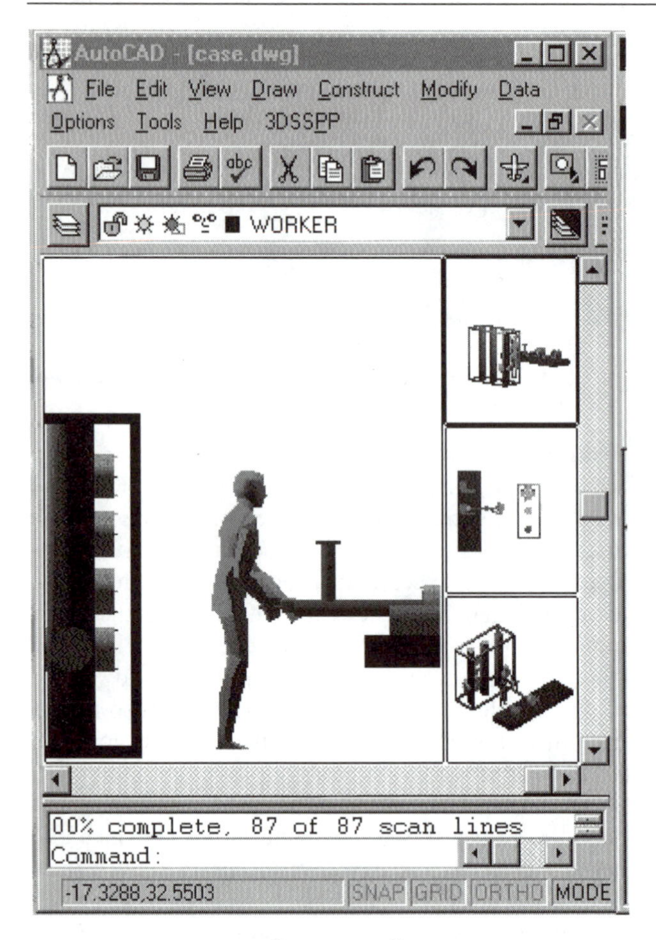

Figure 13.20 The 3DSSPP®/AutoCAD® interface as viewed on the computer screen (from Feyen et al. 1999).

shoulder strength demands would exceed the majority of women's capability at the end of the lift (when the object was over shoulder high).

13.4 SUMMARY

It must be concluded that a valid science exists for identifying excessive peak exertions in many different jobs. The biomechanical strength prediction model and associated software described here provides an effective technology to evaluate jobs and simulate a variety of potential solutions to control overexertion injuries caused by combinations of posture and high hand force requirements.

Work continues at the University of Michigan's Center for Ergonomics to

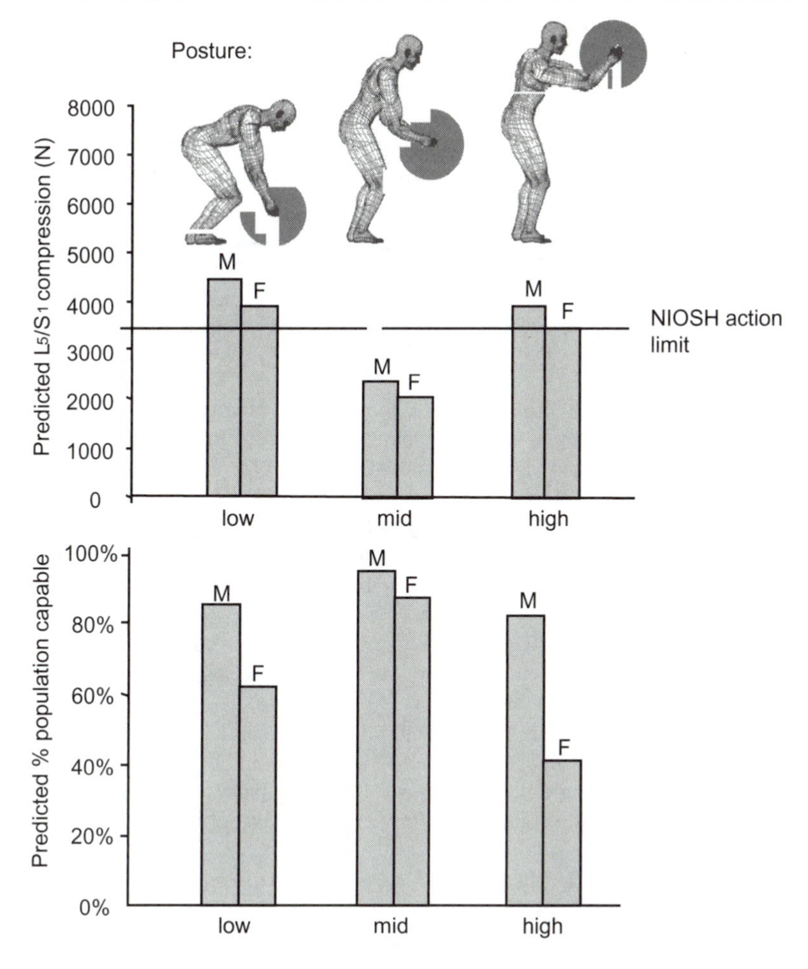

Figure 13.21 Strength model output from analysis of lifting 196 N (44 lb) roll of stock (60 cm dia) into stock feeder at waist and above shoulder height.

allow more dynamic movements to be evaluated using the same logic as described earlier. This work is producing statistical kinematic motion prediction models for a large variety of reaching type motions. By integrating these new motion prediction models with the 3DSSPP® program it will be possible for a workplace designer to insert the hominoid into a virtual workspace, as depicted earlier in Figure 13.20, and then have the model demonstrate realistic motions that would be used to reach and move objects to specific locations. At select points in any simulated motion the designer could ask the 3DSSPP program to perform a biomechanical assessment of the exertion. A first attempt at such a high level analysis is more fully described in Chaffin, Faraway and Zhang (1999).

13.5 REFERENCES

Anderson C, Chaffin D. A biomechanical evaluation of five lifting techniques. *Applied Ergonomics* 1986; 17(1), 2–8.

Bean J, Chaffin D, Schultz A. Biomechanical model calculation of muscle contraction forces: A double linear programming method. *Journal of Biomechanics* 1988; 21(1), 59–66.

Beck D, Chaffin D. An evaluation of inverse kinematics models for posture prediction. In *Proceedings of the Computer Aided Ergonomics and Safety Conference, Tampere,* Finland 1992.

Beck D. Human factors of posture entry into CAD systems, Unpublished doctoral dissertation. University of Michigan, Industrial and Operations Engineering, Ann Arbor, Michigan, 1992.

Byun S. A computer simulation using a multivariate biomechanical posture prediction model for manual materials handling tasks, Unpublished doctoral dissertation, University of Michigan, Industrial and Operations Engineering, Ann Arbor, Michigan, 1991.

Chaffin D, Faraway J, Zhang X. Simulating reach motions, *SAE Human Modeling for Design and Engineering Conference,* The Hague, The Netherlands, 1999.

Chaffin D. Human strength capability and low-back pain. *Journal of Occupational Medicine* 1974; 16(4).

Chaffin D, Erig M. Three-dimensional biomechanical static strength prediction model sensitivity to postural and anthropometric inaccuracies. *IIE Transactions* 1991; 23(3), 215–227.

Chaffin D, Herrin G, Keyserling W. Preemployment strength testingan updated position. *Journal of Occupational Medicine* 1978; 20(6).

Chaffin D, Park K. A longitudinal study of low-back pain as associated with occupational weight-lifting factors. *American Industrial Hygiene Association Journal* 1973; 3, 513–525.

Chaffin DB, Andersson GBJ, Martin BJ. *Occupational Biomechanics,* third edition. John Wiley & Sons Inc., New York, 1999.

Chaffin D, Freivalds A, Evans S. On the validity of an isometric biomechanical model of worker strengths. *IIE Transactions* 1987; 19(3), 280–288.

Evans S, Chaffin D. A method for integrating ergonomic operator information within the manufacturing design process. In *Proceedings of the IIE Conference,* Chicago, IL, December 1985.

Garg A, Chaffin D. A biomechanical computerized simulation of human strength, *AIIE Transactions* 1975; 7(1).

Jäger M. *Biomechanisches Modell des Menschen zue Analyse und Beurteilung der Belastung der Wirbelsäule bei der Handhabung von Lasten.* VDI Verlage, 1987: Reihe 17 Boeechnik N. 33.

Keyserling WM. A computer-aided system to evaluate postural stress in the workplace. *American Industrial Hygiene Association Journal* 1986; 47, 641.

Kilpatrick K. A model for the design of manual work stations, Unpublished doctoral dissertation. University of Michigan, Ann Arbor, Michigan, 1970.

Liu Y, Zhang X, Chaffin D. Perception and visualization of human posture information for computer-aided ergonomic analysis. *Ergonomics* 1997; 40, 818–833.

NIOSH. *Work Practices Guide for Manual Lifting,* DHHS Publication NO. 81-1222. American Industrial Hygiene Assoc., Akron, OH, 1981.

Park K. Computerized simulation model of postures during manual materials handling, Unpublished doctoral dissertation. University of Michigan, Ann Arbor, Michigan, 1973.

Peacock B, Karwowski W. *Automotive Ergonomics,* Taylor & Francis, Bristol, Pennsylvania, 1993.

Snyder R, Chaffin D, Schutz R. *Link System of the Human Torso*, HSRI Report 71-112. Highway Safety Research Institute, and University of Michigan, Ann Arbor, MI and AMRL-TR-71-88, Aerospace Medical Research Laboratories, Ohio, 1972.

Chapter 14

Cost-Benefit of Ergonomics Improvements

Maurice Oxenburgh

14.1 MOTIVATING THE EMPLOYER

Why is it that we still have upper limb and low back disorders related to work? After all, there has been much written about its cost to society, to the insurance industry and to individuals but it is noteworthy that very little has been said about the cost to the individual employer. In many countries there is a direct additional cost to the employer (the workers' compensation insurance premium) if there are injuries but, more often than not, the employer does not carry the entire cost of the injury.

After paying the workers' compensation insurance premiums the employer, typically, does not have to worry any more about injury costs – that now becomes the insurance company's problem. Sure, the premiums may go up but that is just a normal part of running a business. If you are a small company you may never have an injury so injury-related increased premiums are even less of a problem. So why do anything to reduce injuries?

In larger companies that carry their own workers' compensation insurance and/or have their own health, safety and environment departments, there may be motivation to control the incidence of upper limb and low back disorders. In those larger companies that have been successful in controlling injuries it is more often than not due to the efforts of individuals than part to the company system.

So we come back to the question 'how do we motivate the employer to put in place systems that effectively control upper limb and low back disorders?' and is that the right question to pose? Perhaps a better question would be 'how do we encourage employers to count the cost of less-than-optimal workplace conditions?' pointing out that injuries, or even fatigue or tiredness, may be a sign of poorly designed workplaces. Put more bluntly, are there methods to show employers that the profit generated from good workplaces is greater than the profit from poor workplaces? Read on ...

14.2 MEASURING THE COST OF PRODUCTIVITY

To my way of thinking there are methods to show that the profit motive can be used to give better workplaces. First, I think that we occupational health and safety practitioners must get away from trumpeting the idea of injury prevention. It may be the moral situation and one that motivates those of us in the occupational health field but will it motivate employers? I have found that ethical considerations do motivate some managers but that cost considerations, more often than not, will override ethics; we do live in a capitalist/market economy and cost and profit considerations ride high.

A major career block in product or service industry management seems to be that costs for injury prevention are the 'here and now' and benefits are either not counted or are in the future. A manager who spends now may not be in that job in the future to reap the benefits - assuming that the enterprise has a method to measure injuries that do not happen (are prevented); this is an accounting as well as a perceptual block.

Why are benefits not counted? It may be, in part, due to the present accent on the day-by-day (or even hour-by-hour) share price or that profits are only measured in yearly periods. But that is not the entire picture; there is the difficulty of measuring productivity.

It has always seemed strange to me that employers often do not measure productivity or, if they do, it is related to the end-of-the-year overall profit (the 'bottom line') and not to the productive steps within the process. Thus the process is 'hidden' and that makes the task of implementing ergonomics programmes to prevent upper limb and low back disorders more difficult. The present folly of dividing a process into 'cost centres' is an accounting process that allocates costs but not productivity; it is a 'bottom line' accounting process.

However, I am always an optimist and I believe that the productivity benefits accruing from a good ergonomics programme can be measured and measured in a meaningful way. NIVA has organized two seminars to date (Finland, unpublished) to discuss and train occupational health and safety practitioners in economic arguments and calculations. Furthermore, in 1997 there was a conference (The European Conference on Costs and Benefits of Occupational Health) devoted to the cost benefit analysis of occupational health and safety. These are encouraging steps in alerting the occupational health industry that measuring productivity is a valuable tool; a wake-up call that occupational health and safety practitioners must be part of the enterprise and not just an 'add-on'.

In my book (*Increasing Productivity and Profit through Health & Safety*, 1991) I described a cost-benefit analysis method. This method, from the original work of Dr P. Liukkonen of Stockholm University (Liukkonen 1990) is based on the paid hours worked by an employee and the costs of employment (wages), supervision, training and recruitment. Most importantly, the loss of productivity and quality due to poorly designed work places (design, tools, capital investment, organization) is also included in the calculations.

The book's model has since been made into a computer programme (Oxenburgh and Fenestra Software 1999) and this is what I will now describe. The advantage of a computer-generated model is that it handles the data and mathematics and determines the benefits, including productivity increases. As importantly, it acts as a checklist to ask pertinent questions.

Very briefly, a cost benefit analysis takes the workplace situation at a particular point in time. It assumes that productivity is not optimal and that changes or an intervention may be made to improve productivity. The data collected is centred around the people and not the product; the determination is of the workers who produce the goods or services and not of the goods or services. Assumptions are made regarding the costs for implementation of the changes for improvements in the workplace. This is little different from the assumptions made when an engineer wants to introduce new machinery, the travel agent wants extra computers or the grocer/supermarket wants extra storage space. The assumption, even if unwritten, is that the expenditure will result in maintained or extra profit.

Let me take a very simple example to illustrate the value of a cost-benefit analysis although in this case it did not require a computer program to analyse the data (WorkCover 1999). The workplace was the haberdashery section of a department store. There had been some minor back injuries but only sufficient to alert the safety officer that there may be a safety problem. In the section in question the sewing machines were arranged so that the lightest ones (7 kg in weight), and most frequently demonstrated, were on the top shelf of the display stand and could only be reached by standing on a chair. The heaviest ones (18 kg) were well below knee level and had to be lifted onto a table to show the customers.

Minor rearrangement of the shelves and other display stands enabled the lighter machines to be reached safely and the heavier machines did not need lifting at all. The interesting aspect of this example is not solely the reduction in injuries but that the better display of the sewing machines increased sales by 30%.

Although the impetus for the changes came from injuries, the actual re-design came from the saleswomen themselves. In other words, the knowledge for increased productivity was already there but needed some release. It is not necessary to have a computer program to enable this type of work to be done but often some form of checklist is needed to guide the direction of the questions.

This next example is more complex and concerned nurses suffering musculoskeletal injuries from their work in a nursing home (a hospital-like environment for elderly and disabled people). The stimulus to reduce the injury rate was financial; the workers' compensation insurance premiums were increasing and affecting the enterprise's profit.

The concept, to reduce these injuries and eventually insurance premiums, was to reduce lifting to the minimum practicable and this involved considerable capital investment. The type and cost of equipment is illustrated in Table 14.1.

Table 14.1 Type and cost of equipment for the nursing home.

New equipment	40 slings for lifters ($250 each)	$10,000
	3 extra lifters ($2500 each)	$7,500
	20 electric beds ($1600 each)	$32,000
Modify existing equipment	20 lounge chairs on wheels ($500 each)	$10,000
	16 commodes on wheels ($50 each)	$800
	toilet/shower conversions	$4,050
COST OF THE CHANGES		$64,350

Accompanying these physical changes were improved training in patient handling, changes to the patient schedules to avoid stress to the nurses and changes to the nurses' uniforms to be more comfortable.

The benefits that accrued to the nursing home in one year (see Table 14.2) takes into account only the basic savings due to reduced workers' compensation premium from less injuries and not the quality improvements. These quality improvements included the nurses spending more time with each patient: this is not only important for improved patient care, it is important in a competitive, private (non-government) industry. The relatives (decision makers) of the potential patients (clients) for the nursing home can see the quality of care and compare it with competing nursing homes.

Table 14.2 Benefits due to the changes.

Details	*Old system per year*	*New system per year*	*Difference between new and old*
Workers' compensation premium	$324,000	$168,000	$156,000
BENEFIT DUE TO THE CHANGES			$156,000
COST OF THE CHANGES			$ 64,350
Pay back period			5 months

The point I am making here is that management identified the pertinent financial and working condition questions; these led to solutions that reduced injuries and increased productivity and profit.

14.3 CONCLUSION

To return to the question, how does one motivate employers? It is a matter of perception. In the examples above the technical solutions for reduced risk of injury were largely from the employees themselves and/or the solutions were well known to the industry. Although in each of the cases it was the aware employer who took the initiative it still leaves the question how to motivate an employer who ignores injuries or does not take the initiative to reduce costs by workplace improvements.

Even if an employer believes he or she can ignore injuries, no employer can afford to ignore profits. Occupational health and safety practitioners must be part of the profit of a company and hence we need to drop the idea that we are only concerned with injury prevention – we must be seen to be part of the profit generation and not simply a cost. Injury prevention is the tool we use but increases in productivity must be seen to be our aim.

In the second of my cases it was the occupational health and safety officer who determined that the workers' compensation insurance premiums were too high and could be reduced and did the preliminary calculations to show that the exercise would be cost-effective.

In order to be integrated into enterprise systems occupational health and safety practitioners must develop methods directed towards profit increases, although with ethical considerations. The point is that we must measure the effectiveness of our interventions, of ergonomics programmes; our interventions must be measured and shown to be profitable. It will often be a difficult moral balancing act to reduce injuries and to increase company profits, but to do one we may need to do the other.

A model has been presented (*The Productivity Model*, Oxenburgh and Fenestra Software 1999) which calculates worker output in terms related to productivity and profit; most importantly the model acts as a generic checklist to direct the user to identify poor working methods which lead to injury and other unnecessary costs. The checklist assists and encourages the user to look for better and safer ways to do the task with a re-calculation of the productivity and profit (pay-back period) built in to the model. However, motivation of the employer is still an open question; motivation is related to knowledge and perception and our role as occupational health and safety practitioners is to show that good ergonomics is an essential part of the enterprise; it is not an add-on, it is an integral part of the business.

14.4 REFERENCES

Liukkonen P. *Se och räkna på arbetsmiljön* [Count the working environment]. Arbetarskyddsstyrelsen, Stockholm, 1990.

Mossink J, Licher F, eds. *European Conference on Costs and Benefits of Occupational Health.* TNO, Amsterdam, 1997.

Oxenburgh M. *Increasing Productivity and Profit through Health & Safety*. CCH, Sydney, 1991.
Oxenburgh M and Fenestra Software 1999. *The Productivity Model.* Software obtainable from M Oxenburgh by e-mail: maurice_oxenburgh@compuserve.com
WorkCover Authority of NSW. *Linking Productivity to OHS: a quick guide to costs and benefits*. WorkCover Authority of NSW, GPO Box 5368, Sydney, NSW 2001, Australia, 1999.

Chapter 15

Clinical Case Management and Rehabilitation of Workers with Upper Extremity Musculoskeletal Disorders

A Model for Integration of Medical and Ergonomic Approaches

Alfred Franzblau and Thomas J. Armstrong

15.1 INTRODUCTION

This presentation will provide an overview of an ongoing research project which is focused on developing a systematic, multidisciplinary team-based approach to successful rehabilitation of injured workers. The Rehabilitation Engineering Research Center (RERC) based at the University of Michigan Center for Ergonomics is funded for five years by the United States Department of Education, and involves investigators from engineering/ergonomics, physical medicine and rehabilitation, occupational therapy, occupational/organizational psychology, vocational specialists, occupational medicine, and other disciplines. The RERC started in 1998.

The general aim of this project is to maximize the participation of individuals of all ages with disabilities in activities of work and daily living. The specific aim of this project is to develop a Model System for applying ergonomic technologies to accommodate disabled workers. The Model System will include procedures for assessing workers, analysing jobs, identifying accommodation needs and selecting appropriate interventions (ergonomic interventions, work/worksite modifications, assistive technologies, and medical interventions). The Model will be implemented in a database that will be made available to others as a Web-based product. Evaluation of the Model System approach will be used in an iterative fashion for Model refinement. The project, and the Model System, are intended to encompass physical disabilities from all causes (e.g. low back pain, lower extremity problems), but this presentation will focus on the upper extremities.

15.2 BACKGROUND

A person's ability to successfully complete a given task is a function of his or her capacities, the task demands and physical and social environment. Gaps between capacities and demands were fundamental to the process of natural selection. Tools and weapons evolved as ways of overcoming gaps between individual capacities and task demands and contributed to our survival. This early engineering was an important factor in the survival of early man and of our economic and political development, yet many still were excluded because of conditions of birth, disease or injury. The ability of early health care providers to enhance or restore function was very limited. Early assistive technology was limited to a cane or crutch. Until the twentieth century, gaps between capacities and demands or in some cases perceived gaps at best meant surviving in abject poverty.

The twentieth century saw the passage of workers' compensation acts, which required employers to provide for injured workers and helped to foster the development of vocational rehabilitation. In the United States the Social Security Act provided support for persons with insufficient work capacities due to non-work conditions. While these events are important social milestones, many persons were still excluded from work and independent living. With the passage of the Rehabilitation Act of 1973 and the Americans with Disability Act in 1992, providers of goods and services and employers are required to provide reasonable accommodations where gaps exist between individual capacities and job task demands. There is still widespread political debate about the implementation of the Americans with Disability Act. For most people, being unable to fully participate in work means living in poverty. There is a great social and economic need to understand the factors and processes affecting people's ability to participate in society and specifically work. Towards this end, numerous models have been proposed for understanding and facilitating work participation.

Common to all models is juxtaposition of individual traits with those of the job. Also common to all models is the problem that there are so many possible human and job traits that complete enumeration is problematic. As a result, one model may emphasize social factors, while another emphasizes physical factors or one model may emphasize medical factors and another engineering factors. The included factors usually depend on the background of the model designers and the setting in which the model was developed or is to be applied. A model developed for application in one setting may not work in another or may be so general that much work is required to define and validate the important variables. A comprehensive review of all of the models is beyond the scope of this chapter, but it is important to recognize the foundation on which our model is built.

Models can be categorized according to their intended use. Armstrong et al. (1993) proposed a model to explain the relationship between environmental factors, work activities, and internal responses (see Appendix A). The purpose of this model was to provide a vehicle for integrating and interpreting literature from different disciplines, e.g. physiology, biomechanics, psychology, epidemiology, etc. A similar model was recently proposed by a committee of the National Research Council (1998) to integrate and interpret our current knowledge of musculoskeletal disorders and their relationship with work (see Appendix B). While rehabilitation was not the intended application of these models, an understanding of the external exposures and internal responses is necessary for placement of workers who may be at risk of developing or aggravating a medical condition.

A second group of models is used to understand disability in a broad social context. Examples of such models include the ICIDH by the World Health Organization (WHO 1999) and that put forth by the Institute of Medicine (Brandt and Pope 1997) (see Appendices C and D). These models are intended to guide policymakers and the collection of demographic data for evaluating programs and policy. These models generally are too general for individual case management, but the concepts can help a professional develop strategies and identify important factors.

A third group of models is used to gather specific worker and job information for case management and can be characterized as 'end-user' models. While these models are conceptually similar to the aforementioned models, they can provide information immediately useful to the practitioner. An early U.S. example of such a model is that proposed by Bridges in 1946. Bridges proposed lists of job and corresponding worker attributes. The user collected the relevant information about the job and job applicant to identify gaps between job demands and applicant abilities. Similar systems have been proposed by others, perhaps the most recent is ERTOMIS (Nieuwenhuijesen 1990). While these systems are helpful in many situations, they lack sensitivity and specificity.

Disability can be determined by: (1) comparing an assessment of a worker's abilities and performance with requirements for a given task before they are placed, (2) comparing a worker's performance with the task requirements after they are placed or (3) monitoring morbidity patterns after they are placed. The first approach is most ideal in that it allows health care providers to determine if medical, technological, or work interventions are required and how best to implement them. This approach can minimize the time cost required to successfully place workers and avoids exposing workers to activities that may be unsafe or unhealthy. It also avoids putting workers into situations where they are likely to fail and the subsequent psychological effects. Unfortunately, it is the second and third approaches that are used most frequently. As a result, many healthcare providers and employers often rely on appearances and impressions, rather than more objective job task assessments, to make worker placement decisions. As a result they may not even attempt to place workers with conspicuous impairments. In addition, it is often found after the fact from subsequent medical complaints and workers' compensation reports that the work requirements are excessive for workers who appear to be able to perform a job.

Rehabilitation in the workplace is concerned with assessment and restoration of human performance so that an individual can perform a given task with minimal risk to himself or herself and his or her co-workers. Ergonomics is concerned with assessment and design of work requirements so that a given task can be performed by a given individual without risk of injury to him or herself or his or her co-workers. Ergonomics can enhance the rehabilitation process and maximize the likelihood of individuals with all levels of physical and mental abilities gaining meaningful employment while minimizing their risk of injury or illness or in some cases reinjury or reillness. A model of this process is summarized in Appendix E. The proposed model is conceptually similar to models described above. The main difference is in its implementation. We propose a hierarchical approach that begins with an overview of the job to determine required work tasks and demands. This is very similar to the 'Task Inventory' method described by McCormich (1982). This makes it possible to minimize the collection of unnecessary job and worker data while maximizing the collection of data about essential job functions.

15.3 THE MODEL SYSTEM

Many of the elements of the model shown in Appendix E have already been implemented; however they are seldom implemented in a systematic manner. For example, a Quantitative Functional Assessment (QFA) (Haig 1998) has been applied to assess worker performance. While the QFA provides important information, comparable work requirement data are often not available. As a result, many healthcare providers and employers will be excessively conservative in placement of workers with conspicuous impairments or workers are placed in jobs in which they are likely to fail. This uncertainty of success discourages many employers from investing in workplace accommodations.

Similarly, systems have been developed for analysis of work requirements and stresses. Examples include: the motion inventory analysis system, various checklists, systems for observing and categorizing work activities, and instrumental based systems that record work posture and muscle activity. Most of these have not considered worker assessments or risk of musculoskeletal disorders.

A hallmark of ergonomics and rehabilitation is teamwork. From Halstead's initial critical review (1976) through the comments of Diller (1990) in this decade, great emphasis has been placed on the benefits of interdisciplinary and even transdisciplinary teamwork. Most of the research which has been done addresses inpatient or subacute rehabilitation. Most of the outcomes have to do with survival, length of stay, cost, and general disability. Outside of the spine literature (e.g. Hazard 1995), we have found very little except general discussion on the effect of team intervention on work, and little on the integration of ergonomic expertise into the rehabilitation team.

In the past, vocational rehabilitation specialists (but almost never ergonomists), have been an integral part of the medical rehabilitation team. As hospital length of stay decreases and health insurers commonly exclude vocational interventions, the medical rehabilitation team and the vocational rehabilitation team are increasingly separated– the first funded by health insurance, and the second only invoked after the patient has reached a healing plateau– after therapies are often complete. The opportunity for ergonomic interventions to be integrated with medical interventions is slim.

15.4 CASE SELECTION

Clinical experience tells us there is an unknown, but likely great benefit from ergonomic advice within a medical rehabilitation team at the time when patients are in the planning stages for return to work. Furthermore, recent data suggest that the prevalence of symptoms and/or diagnosable disorders of the upper extremities among active workers is substantial. In a study of 352 factory workers that utilized standardized clinical tools for assessing the upper extremities, 129 (37%) reported symptoms in the wrists/hands/fingers, 35 (10%) had tendinitis in the upper extremities based on symptoms and physical examination findings, and 19 (5.6%) had carpal tunnel syndrome based on symptoms and sensory nerve conduction tests (Latko et al. 1999). Thus, ergonomic interventions can be targeted for active workers with the goal of maintaining productive employment and reducing symptoms and/or injury, in addition to aiding in the rehabilitation of injured workers who have been disabled from employment. The point at which ergonomic intervention is most effective can be surmised from an understanding of the process of rehabilitation which different groups of patients undergo.

The overall aim of this work is to develop a comprehensive approach involving rehabilitation medicine and ergonomics to facilitate placement of workers and to prevent development or worsening of musculoskeletal illnesses and injuries. Our approach is to develop a Model System to combine ergonomic interventions, work/worksite modifications, assistive technologies, and medical interventions to accommodate disabled workers. The RERC Core Projects are focused on this goal and include four closely related projects dealing with: worker ability/performance assessment; job/worksite analysis, identification of accommodation needs together with recommendations for interventions, and an action research methodology for applying, evaluating, and refining the Model System. All of the Core Projects will work together in close collaboration in order to integrate these components into the Model System. Development of the Model System will entail the broadening of the rehabilitation team to include ergonomists. The actual form of the Model System will be a database structure which includes information on a broad range of interventions and case examples. This process will culminate with a Web-based Model System that can be used by rehabilitation professionals, employers, consumers, and organizations.

15.5 CASE STUDIES

In this presentation the Model System will be illustrated via a limited number of examples of cases and jobs. The idea is to illustrate how detailed and objective ergonomic information related to the physical stresses of jobs can assist in proper placement of injured workers in order to optimize recovery and to enhance continued productivity. We have selected four jobs, and four cases involving the upper extremities, to be presented and discussed. The medical details of each case are summarized below, together with the descriptions of a sample of jobs.

Case 1: 32-year-old right-handed female with bilateral forearm pain
Medical history: non-contributory
Physical examination: no abnormal findings
Laboratory tests: MRI of neck and forearms normal; tests for muscle inflammation were normal
Prior treatments: NSAIDs, ice, exercise/physical therapy
Job: Hospital billing clerk (job A)

Case 2: 28-year-old right-handed male with right shoulder pain
Medical history: non-contributory
Physical examination: tender over supraspinatus tendon; pain over supraspinatus tendon with active abduction; no crepitus or muscle weakness
Laboratory tests: none
Prior treatments: none except occasional NSAIDs Job: spark plug manufacturing assembly line (job D)

Case 3: 48-year-old right-handed male with right lateral elbow pain
Medical history: none, except likes to golf
Physical examination: point tenderness just distal to right lateral epicondyle; pain with resisted extension of right wrist/fingers; no locking/clicking of elbow and no muscle weakness

Laboratory tests: none Prior treatments: none
Job: spark plug manufacturing assembly line (job D)

Case 4: 36-year-old right-handed female with intermittent burning, numbness, soreness, and tingling in both wrists, hands, and fingers
Medical history: thyroidectomy, on replacement therapy; history of tendinitis in left wrist; BMI = 31.5 kg/m^2
Physical examination: no tenderness; no crepitus or locking/clicking with active or resisted motion in any joints in the wrists, hands, or fingers; strength in the opponens pollicis is normal bilaterally; no atrophy; Phalen's and carpal compression tests are positive bilaterally; twopoint discrimination is normal in digit 2 bilaterally
Laboratory tests: median sensory latencies across carpal canal abnormal; corresponding ulnar latencies normal
Prior treatments: NSAIDs, rest, splints, physical therapy, restricted duty Job: office chair upholsterer (not listed in Table of Jobs)

Table 15.1 Jobs and distinguishing attributes.

Office/Clerical/Medical Billing (A)	*Office/Clerical/Transcriptions (B)*
Workstation/equipment: • Computer workstation • User adjustable keyboard tray • Documents located on separate tables on right and left • Phone with handset • Conference room	Workstation/equipment: • Computer workstation • User adjustable keyboard tray • Tape player with foot control and headset
Tasks/Methods: • Keyboard 10% • Mouse 4% • View screen 8% • Writing 1% • View charts 25% • View refs 6% • Phone 4% • Away 34%	Tasks/Methods: • Load/unload tape player • Transcribe tapes 80% • Look up words • Proof read documents
Repetition: Low, some prolonged static exertions	Repetition: High fingers/wrist, prolonged exertions elbows/shoulders/neck/back
Force: Medium peaks to handle charts and docs	Force: Low–medium peaks keying
Contact: Low	Contact: Low–high average depending on workstation adjustment

Office/Clerical/Medical Billing (A)		*Office/Clerical/Transcriptions (B)*	
Posture:	Medium–high peak elbow for keying; medium–average for neck to view charts and documents	Posture:	Low–high wrist and shoulder depending on workstation adjustment; medium–high elbow keying

Manufacturing/Assembly/ Auto transmissions (C)		*Manufacturing/Assembly/Auto sparkplugs (D)*	
Workstation: • Assembly on indexing line • Assembly rotated to work position • 150–160 units/hr • Seated		Workstation: • Moving assembly line 1 – in • Moving assembly line 2 – out • Machine paced • Sit or stand	
Tasks/Methods (monotask): • Advance line and rotate assembly to work position • Adjust fixture/assembly • Get and install small parts		Task/Methods (monotask): • Get parts from line 1 – both hands • Inspect • Position parts in holders on line 2	
Repetition:	Medium	Repetition:	Very high
Force:	High peaks to install parts	Force:	Low
Contact:	Medium peak stresses on fingers and palm	Contact:	Low
Posture:	High peak stress on wrist, elbows and shoulders to get and assemble parts	Posture:	Medium–high peak stresses on wrist and elbow to get parts; medium–high peak stresses on wrist, elbows and shoulders to get parts

15.6 JOBS

15.6.1 Job A: Office/Clerical/Medical billing

This job entails multiple intermittent tasks. Tasks include: locating billing information from documents, looking up billing rules from manuals and computer, entering data in computer, making and handling phone inquiries and attending meetings. Repetition, force exertions, contact stresses and postures stresses are generally low, but could be higher depending on adjustment of work equipment and individual work method. Also, there could be irregular periods of prolonged exertions with certain work schedules. Certain interventions might help to reduce stresses and accommodate persons with extremely limited tolerance for handling documents and computer work:

- careful adjustment of the keyboard, monitor, documents, and seating;
- use document holders;

- provide equipment maintenance;
- provide structured breaks;
- provide worker training.

15.6.2 Job B: Office/Clerical/Transcriptions

This job mainly entails entering alphanumeric text into a computer from audio tapes. The tape position and speed are controlled via a foot control. The job is interrupted every ten to thirty minutes to load and change tapes and irregularly to look up words. Ergonomic exposures include high repetition of the wrist and fingers to enter data and prolonged exposures of the upper and lower body to operate the foot control and maintain keying position. Exposures to force, contact stresses and posture stresses are generally low, but may be high depending on how the worker performs the job and how they adjust the workstation. Certain interventions may help to reduce the stresses and accommodate people with limited capacity for prolonged and highly repetitive keyboard work; however, these interventions will not eliminate the high repetition inherent in this work, which could still be a problem for some workers:

- use ergonomic keyboard (low key activation force, shaped to reduce wrist deviation);
- use careful adjustment of the keyboard, monitor, documents, and seating;
- use document holders;
- use structured breaks;
- provide equipment maintenance;
- provide worker training.

15.6.3 Job C: Manufacturing/Assembly/Auto transmissions

This is a monotask job that entails getting and inserting parts on an assembly line. The parts all weigh less than 1 N, but high exertions may be required to seat them. Medium forces are required to position the fixture in work position. The average work pace is medium, but there may be periods of idleness followed by periods of high repetition due to line irregularities. The part stops in front of the worker, which eliminates reaching up and down the line; however the seated work position results in extreme forward reaching to insert parts into the assembly. Extreme reaching side to side along with extreme forearm rotation and wrist flexion is required to get parts from the seated position. Certain interventions might help to accommodate persons with low reach capacity; however, this job will still entail medium repetition and forceful peak exertions:

- use auto-rotation of the fixture/assembly into work position;
- locate the parts in trays close to their point of use to reduce reaching;
- use adjustable cantilever seating to permit the worker to get as close as safely possible to the work;
- elevate the fixture so that the job can be performed in a stand/sit position;
- select gloves that optimize fit and grip;
- educate worker about positioning work equipment.

15.6.4 Job D Manufacturing/Assembly/small parts

This is a highly repetition monotask job that entails low forces, and medium wrist and shoulder posture stresses. Certain interventions might be implemented to reduce some of the handling and reaching associated with the posture stress and high repetition:

- use rollers to orient parts so that they can be inspected without handling them;
- arrange the lines so that parts can be slid from one line to the other;
- adjustable cantilever seating conducive to a sit–stand posture so that worker can get closer to work and reduce reaching.

15.7 CASES

15.7.1 Case 1: 32-year-old right-handed female with bilateral forearm pain

A 32-year-old right-handed female presents with bilateral forearm pain, worse on the volar aspects. She has no other chronic medical problems, and her BMI is 26 kg/m^2. The pain had gradual onset starting in February of 1998, and was noted by the patient to be worse after working 1–2 hours. There was no history of trauma. She does not engage in any conspicuous avocational activities which might cause or aggravate her forearm pain. Her physical examination has been consistently normal, with no focal findings. Laboratory tests for muscle inflammation (sedimentation rate, aldolase, creatine phosphokinase) were unremarkable. Because of the patient's continuing discomfort and anxiety about her pain and prognosis, she underwent a cervical MRI and MRI of both forearms. These studies were normal. She has been treated with nonsteroidal antiinflammatory drugs (NSAIDs), occasional ice, structured exercise and physical therapy, and there have been limited modifications to her job station. She has remained fully employed as a medical records clerk in a hospital billing department.

Workers like Case 1 (i.e. those with symptoms but no abnormal findings on clinical examination or laboratory tests) represent a major fraction of the working population (Latko et al. 1999). The symptoms can vary from being a minor nuisance, to, in rare cases, developing into debilitating pain syndromes. And, it is likely that as the average age of the workforce increases, this profile will become even more common. A key feature of clinical management of this patient/worker is to gain her confidence, and to reassure her that while her pain is 'real', there is no evidence that she is damaging herself by continuing to work. As noted above, she was actually working as a clerk in the medical billing department of a hospital (see Job A, Table of Jobs), which is the least stressful of the jobs chosen for this discussion. Job A incorporates a variety of tasks, with varying amounts of repetition and force, but overall, the peak and average physical stresses are low. In fact, Job A represents the sort of 'limited duty' job that would be suitable for almost any worker/patient with problems that limit the amount of physical stress to which she should be exposed. Hence, it is unlikely that the physical stresses related to Job A were making a significant contribution to her symptoms (it is possible that other types of stress, such as psychosocial stresses, might be significant, but such factors and related interventions, while important, are beyond the scope of this project). Given the already low degree of physical stress of Job A, it is unlikely that ergonomic interventions will dramatically reduce the peak or average amount of physical stress, and so it is unlikely that ergonomic interventions will greatly enhance this worker's ability to perform her job while simultaneously reducing her symptoms. Nevertheless, there are a number of relatively low cost interventions which individually or taken together might be beneficial, such as proper adjustment of existing equipment; document holders, or a structured break schedule. The low cost of such interventions makes it possible to justify them, despite the reduced likelihood of success in this case.

Assigning this worker to jobs such as those illustrated by Jobs B, C, or D would be more problematic (see Table of Jobs). Studies have shown that jobs with greater physical stresses are associated with greater prevalences of upper extremity symptoms and/or signs. In the absence of any specific signs or symptom patterns suggestive of an incipient disorder, there is no basis for restricting her from such jobs. However, if this worker were reassigned to a more stressful job careful monitoring for development of an upper extremity disorder is advisable.

15.7.2 Case 2: 28-year-old right-handed male with right shoulder pain

A 28-year-old right-handed male presents with intermittent right shoulder pain for about one year. There is no other pertinent medical history, except for surgical excision of a ganglion in the right wrist two years earlier, and that he takes Zantac for symptoms of gastric reflux. His BMI is 30.3 kg/m^2. The right shoulder pain is exacerbated by work. The pain does not radiate down the arm or to/from the neck, nor are there symptoms elsewhere in the limb (e.g. elbow, forearm, wrist, or fingers). There is no history of local trauma, nor does he engage in any conspicuous avocational activities which might cause or aggravate right shoulder pain. Typically he experiences one to three episodes per week, each lasting a few hours. He has not sought to change jobs. He has not sought any prior medical treatment for this problem, though he has self-medicated with occasional over-the-counter nonsteroidal antiinflammatory medications with limited relief. On physical examination, the shoulder was tender over the supraspinatus tendon. There was pain in the shoulder with active abduction, particularly over the supraspinatus tendon. Resisted internal and external rotation, and resisted extension/flexion of the right shoulder did not produce pain. There was no crepitus or muscle weakness. He has remained employed in an assembly line job manufacturing spark plugs.

Case 2 represents an example of a worker with a specific disorder, most likely a mild supraspinatus tendinitis, who is engaged in a task which caused, or is exacerbating his condition (see Job D, Table of Jobs). It is not the focus of this presentation to review potential medical interventions for such conditions, and so they will not be discussed. While the major stresses of Job D involve the hands and wrists, the job also requires repeated reaching out to grasp parts, thus repeatedly stressing the injured part. At the very least continued performance of this job will likely retard the recovery of this worker's shoulder disorder (regardless of medical interventions) if it did not materially contribute to, or exacerbate this condition. Therefore, a temporary job reassignment is advisable. This worker would be particularly ill-suited for Job C due to the high degree of shoulder stress involved in Job C. Job A, or similar office duties, would be an excellent choice for a light duty assignment for this worker in order to allow his shoulder condition to recover maximally. However, having had tendinitis, he is at increased risk of developing it again in the same distribution, and so, regardless of which job he is ultimately assigned to, he should be monitored periodically to assess if his condition is stable, or deteriorating. Job B may also be an acceptable light duty assignment for such a worker (of course, depending on skills and ability to perform essential job tasks). The only caveat related to Job B is that as a monotask job, there is a greater likelihood of muscle strain due to the constrained, static posture of performing medical transcription. It would be prudent to provide a well-designed work station that includes arm rests so as to minimize static postural stress on the shoulders. It is also essential that the work station be adjusted properly.

15.7.3 Case 3: 48-year-old right-handed male with right lateral elbow pain

A 48-year-old right-handed male presents with right lateral elbow pain for the last 10 months. There is no other pertinent medical history. His BMI is 27.6 kg/m^2. The pain in the right elbow waxes and wanes, but it has been present to some degree almost constantly for the last few months. According to the patient, the pain is exacerbated by grasping handfuls of boxes, though he has not sought to change jobs. There are no other symptoms elsewhere in the upper extremities. There is no history of local trauma. He likes to play golf 'several times per week, weather permitting'. He has not sought any prior medical treatment for this problem, and he has not self-medicated for this problem. On physical examination, there is point tenderness just distal to the lateral epicondyle of the right elbow. The pain in the right lateral elbow is worsened with resisted extension of the right wrist and/or right fingers. There is no locking or clicking in the right elbow, and there is no muscle weakness. He has remained employed in an assembly line job manufacturing spark plugs.

Case 3 is similar in many respects to case 2. This worker is engaged in a job (Job D), and possibly avocational activities (golf), which have probably caused or exacerbated his condition. Furthermore, despite optimal medical interventions, continued performance of this job and/or his avocational activities will likely delay or even prevent recovery. Therefore, in addition to medical interventions, a temporary job re-assignment would be indicated, along with modification of avocational activities. The justification for possibly reassigning this worker to Jobs A or B (or similar jobs) is similar to that presented for Case 2. And, as with Case 2, job reassignment should be accompanied by medical monitoring to assess recovery and continued job compatibility.

15.7.4 Case 4: 36-year-old right-handed female with symptoms of CTS

A 36-year-old right-handed female presents with intermittent burning, numbness, soreness, and tingling in both wrists, hands, and fingers. Her medical history includes a thyroidectomy at age 26, with subsequent thyroid hormone replacement. Laboratory tests indicate that she has been euthyroid. Two years ago she was diagnosed with tendinitis in the left wrist. This condition was treated with nonsteroidal antiinflammatory medications, rest, splints, and some physical therapy. Eventually, she was put on restricted duty for four months at work until about one month ago. She is not, and has not been pregnant recently, and she does not use birth control pills. Her BMI is 31.5 kg/m^2. She cannot precisely date the onset of her current symptoms, but in the last year she estimates having had two to three significant episodes per week, including nocturnal awakening due to symptoms. On a hand diagram, she localizes the symptoms to include all five fingers in both hands. On examination, there is no tenderness, and no pain, crepitus, or locking/clicking with active or resisted motion in any joints in the wrists, hands, or fingers. Muscle strength in the opponens pollicis is normal bilaterally, and there is no atrophy. Phalen's test and the carpal compression test are positive in both wrists; Tinel's test is negative bilaterally. Two-point discrimination is normal in digit 2 bilaterally. Sensory nerve conduction velocity tests (14 cm antidromic stimulation involving digits 2 and 5) show delayed conduction in the median nerve across the carpal canal (right median sensory peak latency = 4.4 milliseconds; and left median sensory peak latency = 5.8 milliseconds; peak latencies of the corresponding ulnar nerves were normal). She remains employed as a chair upholsterer in an office furniture manufacturing factory.

Case 4 is a case of bilateral carpal tunnel syndrome (CTS) based on the history and symptoms, physical examination findings, and electrodiagnostic test results. The history is particularly interesting in that she had been diagnosed with wrist tendinitis two years earlier. While it is possible that tendinitis may have been present at that time, it is also possible that her symptoms were related to CTS. In many respects the treatments she received for tendinitis overlap with many of the conservative therapies for CTS, and so she may have been partially treated for CTS. This particular worker was engaged in a job that is not included among those analysed in detail for this presentation. However, furniture upholstering is among the most forceful and repetitive jobs we have ever studied, and therefore would qualify as a significant contributing factor in this worker's development of CTS. Clearly, she requires a job reassignment (and, in fact, she was on light duty for a number of months), whether temporary or permanent depends on other factors to be discussed.

As noted above, it is not the focus of this presentation to present or evaluate specific medical interventions for musculoskeletal disorders. Rather, the focus is on possible ergonomic interventions and selection of jobs. However, this case illustrates how choices or timing of ergonomic interventions may be tied to specific medical treatments and their outcomes. CTS can be treated conservatively with a variety of modalities. However, in many cases surgical treatment is indicated, frequently following an attempt at conservative therapy. In this case either approach might be justified. If she and her physician elect to pursue conservative medical therapy, then a light duty assignment would be indicated during the recovery period. Among the jobs described in the Jobs Table, Job A would probably be suitable, while Jobs C and D, which involve considerable hand/wrist repetitive movements, would be unsuitable. Job B, a monotask involving medical transcription, would be a debatable choice. Whether constant typing represents a risk factor for CTS is beyond the scope of this discussion, but the static exertion of the wrists associated with typing (i.e. extension and ulnar deviation) are more problematic (ideally, the work station would be adjustable, and designed to minimize these stresses). Again, this highlights the need for medical monitoring of workers after job placement in order to assess progress. Finally, assuming that conservative medical therapy, with job re-assignment, are successful at treating her CTS, or controlling her symptoms, then there is the decision of whether she can return to unrestricted duty (specifically, her original job). If she developed CTS while performing her original job, then if she is reassigned to the same task it would be likely that her problem will recur. Hence, while she may be able to safely perform jobs with greater physical stress than Job A, it is likely that she would not be able to return to her original job (furniture upholstering), or similar jobs that involve high stress tasks. Again, medical monitoring of this worker's condition by the plant health personnel is important.

At some point, this worker and her physician may elect to pursue surgical treatment for her condition. The question of whether workers with CTS who undergo successful surgical treatment can eventually return to the same jobs is controversial. Some surgeons recommend permanent job reassignment, while others contend that the underlying pathophysiology is changed by the surgery, thus permitting continued performance of stressful jobs after successful surgical therapy. Available data suggest that workers who undergo surgery for CTS tend to fair better than workers who are not treated surgically for up to 30 months following surgery (Katz et al. 1998). In particular, they are more likely to return to work, to have fewer symptoms, and to be more satisfied (Katz et al. 1998). However, what is not clear from this study, or any other studies which have followed workers treated surgically for CTS, is how they have faired if they returned to the same jobs which may have caused and/or exacerbated their CTS. In the study by Katz et al. job 'exposure' was assessed on the basis of job titles, and workers' self-report; there was no independent assessment of job exposures. If workers returned to work after treatment for CTS, whether surgical or conservative, it is unclear if they returned to the same employer, or the same job, or whether there may have been a modification to the job to reduce physical stress. Thus, the study by Katz et al. does not directly address this important question, nor are we aware of any empirical studies which do.

15.8 SUMMARY

These cases and jobs are intended to illustrate the application and potential benefits of the rehabilitation process encompassed by the model shown in Figure 1. Incorporation of detailed ergonomic information into decisions regarding job compatibility and/or job modification can lead to a better match between worker and job attributes, and therefore more successful job placement. The RERC will continue to collect data on jobs and cases, and to better refine the process and model. Ultimately, it is one of the goals of the project to development a product that could be accessed via the web. The product would be designed to serve as a tool to assist practitioners in job placement/job modification decisions by providing a spectrum of jobs and cases which illustrate a wide range of worker and job attributes.

15.9 REFERENCES

Armstrong T, Buckle P, Fine L, Hagberg M, Jonsson B, Kilbom A, Kourinka IA, Silverstein B, Sjogaard G, Viikari-Juntura E. A conceptual model for work-related neck and upper limb disorders. *Scand J. Work Environ Health* 1993; 19(2), 74–84.

Brandt E, Pope A, eds. *Enabling America: Assessing the Role of Rehabilitation Science and Engineering*. Committee on Assessing Rehabilitation Sciences, http://www.nap.edu, National Academy Press, Washington, DC, 1997.

Bridges C. *Job placement of the physically handicapped*. McGraw-Hill Book Company, New York, 1946.

Diller L. Fostering the interdisciplinary team: fostering research in a society in transition. *Arch Phys Med Rehabil* 1990; 71, 275–278.

Haig AJ. Rehabilitation of injured workers. In *Occupational Rehabilitation*, Kaplanski, ed., 1998 (in press).

Halstead LS. Team care in chronic illness: a critical review of the literature of the past 25 years. *Arch Phys Med Rehabil* 1976; 57, 507–511.

Hazard RG. Spine update. Functional restoration. *Spine* 1995; 20(21), 2345–2348.

Katz JN, Keller RB, Simmons BP, Rogers WD, Bessette L, Fossel AH, Mooney NA. Maine carpal tunnel study: Outcomes of operative and nonoperative therapy for carpal tunnel syndrome in a community-based cohort. *J Hand Surg* 1998; 23A, 697–710.

Latko WA, Armstrong TJ, Franzblau A, Ulin SS, Werner RA, Albers JW. A cross-sectional study of the relationship between repetitive work and the prevalence of upper limb musculoskeletal disorders. *Am J Ind Med* 1999; 36: 248–259.

McCormick E. Job and task analysis. Chapter 2.4, in *Handbook of Industrial Engineering*, G Salvendy, ed., John Wiley, New York, 1982: 2.4.4–2.4.6.

National Research Council. *Work-Related Musculoskeletal Disorders: A Review of the Evidence*. National Academy Press, 2101 Constitution Avenue N.W., Washington, D.C. 20418, 1998. http://www.nap.edu

Nieuwenhuijesen E. The ERTOMIS assessment method: An innovative job placement strategy. *Forging Linkages*, M Berkowitz, ed. Rehabilitation International, New York, NY, January 1990: 121–156.

WHO. *International Classification of Functioning and Disability Beta-2 Assessment, Classification and Epidemiology Group*, WHO/HSC/ACE/99.2. World Health Organization, Geneva, Switzerland, July 1999.

Appendix A – A ‘Model’

from: Armstrong et al. (1993)

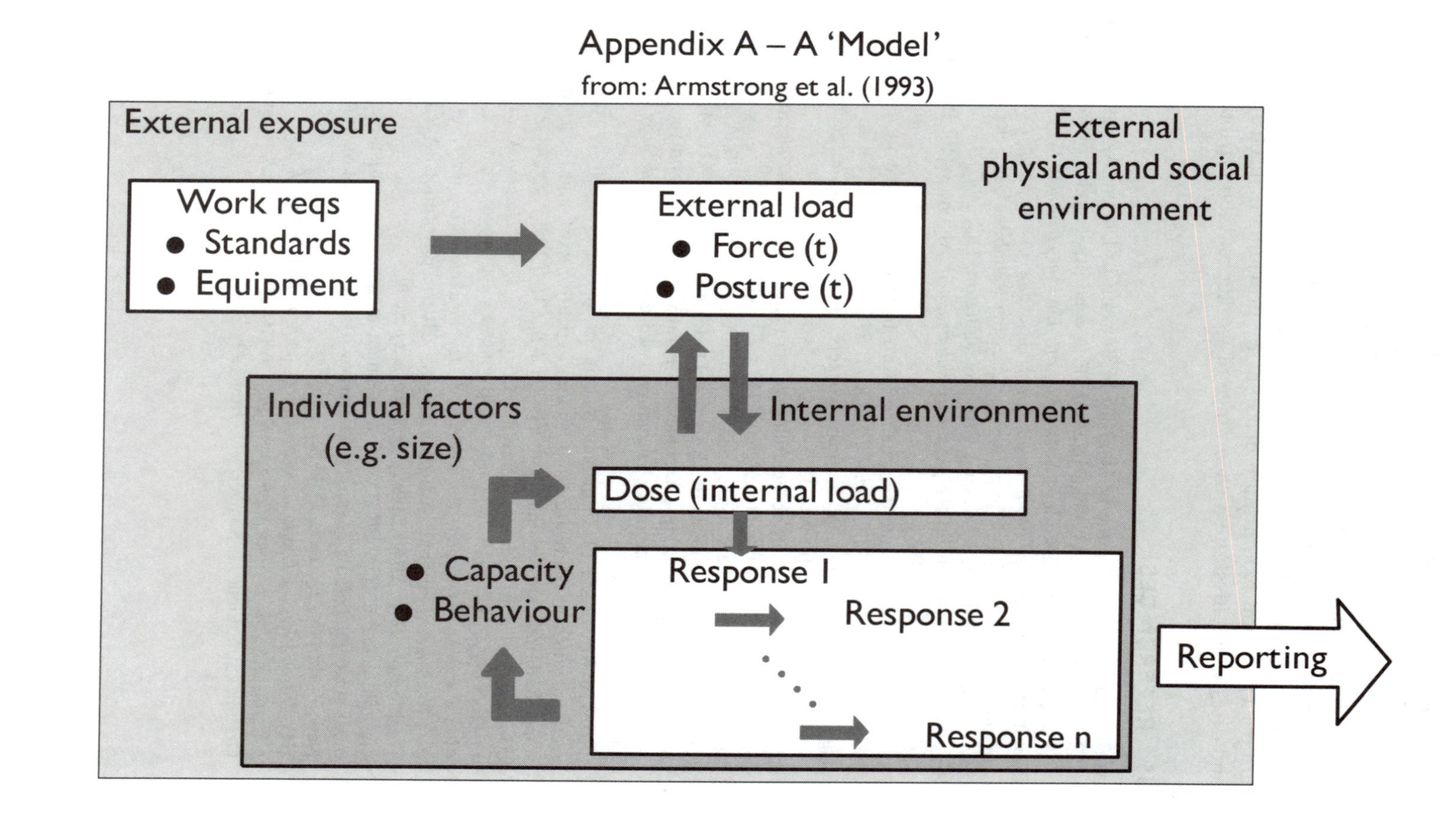

Appendix B – NAS 1998

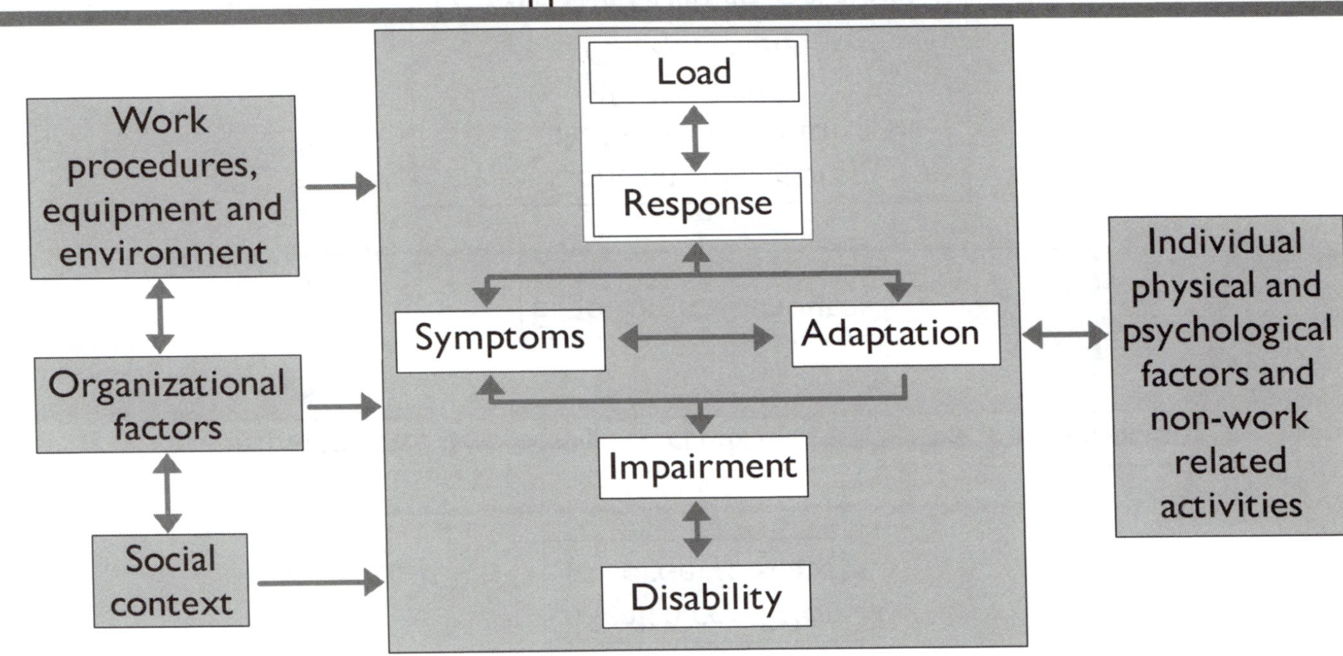

National Research Council. *Work-Related Musculoskeletal Disorders: A Review of the Evidence*. National Research Council, National Academy Press, 2101 Constitution Avenue N.W., Washington, DC 20418, 1998. http://www.nap.edu

Appendix C – ICIDH-2

(WHO 1999)

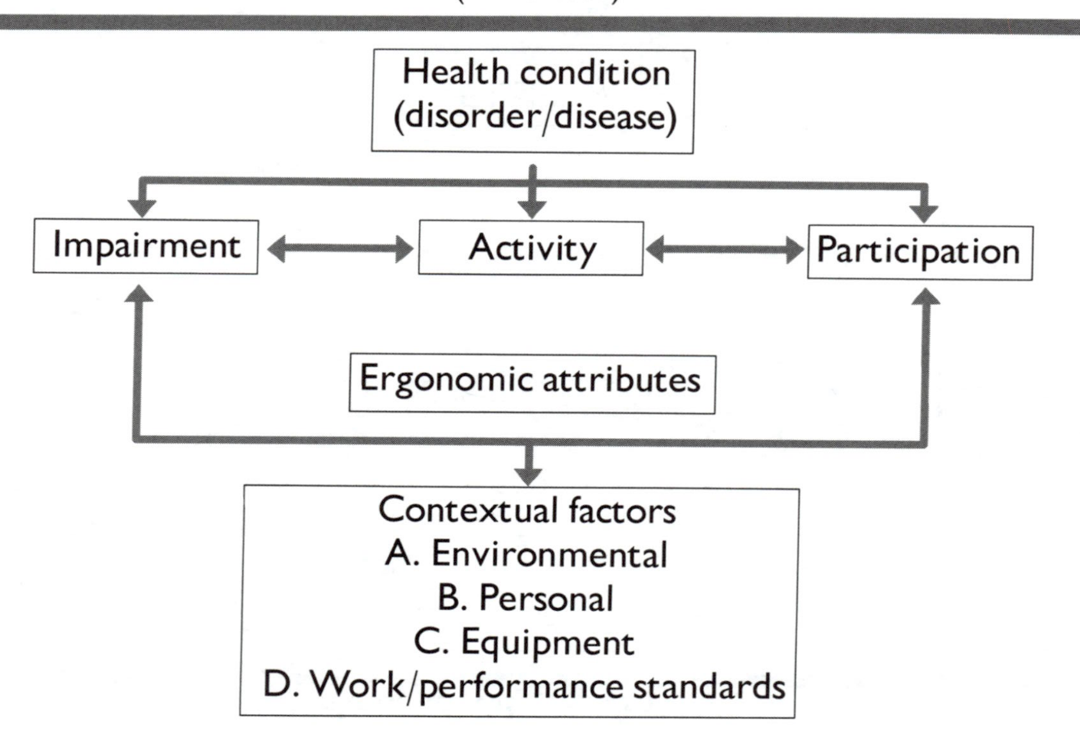

Appendix D – Institute of Medicine Disability Model

Pathology	*Impairment*	*Functional limitation*	*Disability*
Interruption or interference of normal bodily processes or structures	Loss and/or abnormality of mental, emotional, physiological, or anatomical structure or function; includes all losses or abnormalities, not just those attributable to active	Restriction or lack of ability to perform an action or activity in the manner or within the range considered normal that results from impairment	Inability or limitation in performing socially defined activities and roles expected of individuals within a social and physical context
Level of reference			
Cells and tissues	Organs and organ systems	Organism – action of activity performance (consistent with the purpose or function of the organ or organ system)	Society – task performance within the social and cultural context
Example			
Denervated muscle in arm due to trauma	Atrophy of muscle	Cannot pull with arm	Change of job; can no longer swim recreationally

Brandt E, Pope A, ed. *Enabling America: Assessing the Role of Rehabilitation Science and Engineering*. Committee on Assessing Rehabilitation Sciences. http://www.nap.edu. National Academy Press, Washington, DC, 1997.

Appendix E – Model System

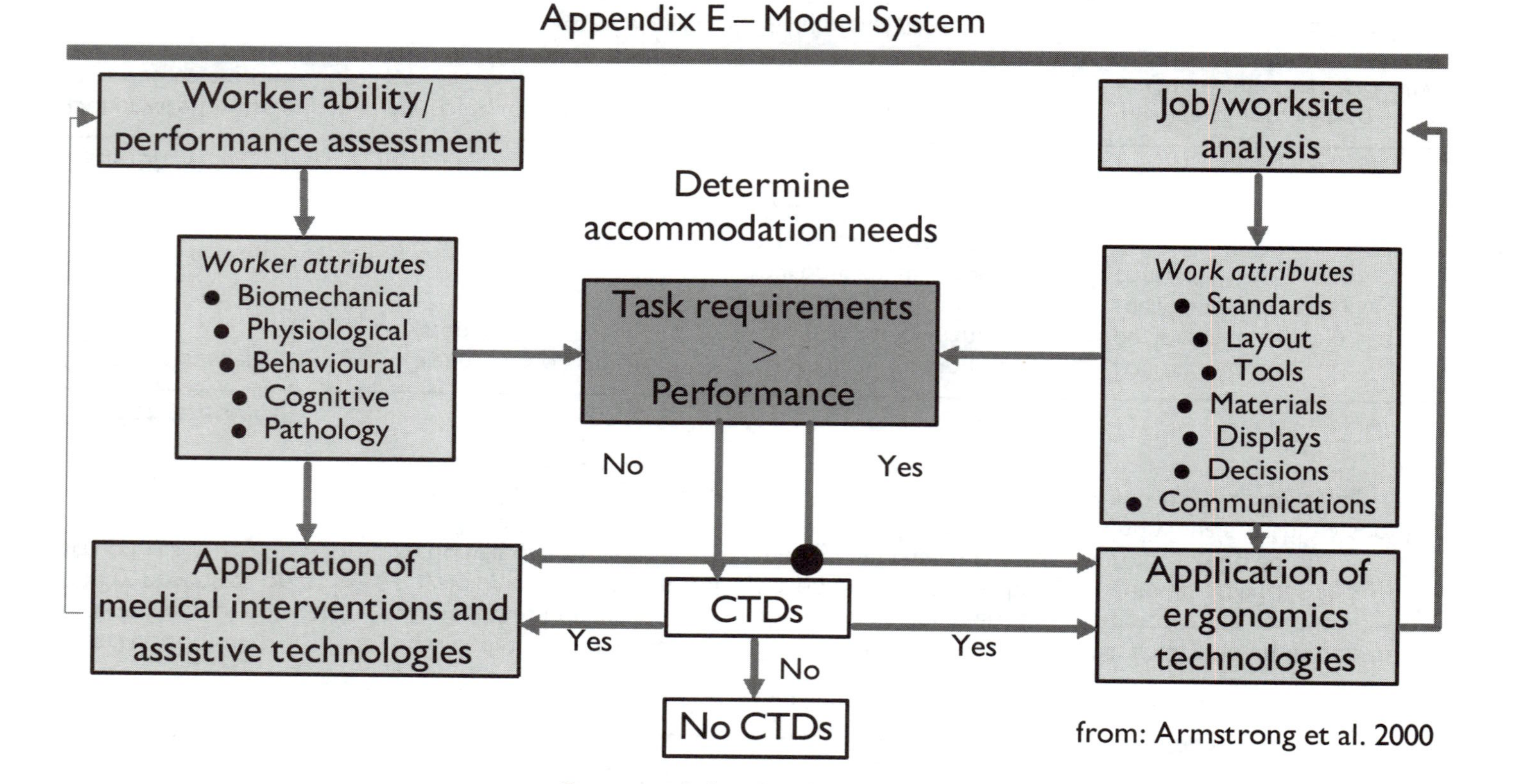

Chapter 16

Experiences in Redesigning Jobs Featuring Repetitive Tasks

Daniela Colombini, Enrico Occhipinti, Renzo Bergamaschi and Candido Girola

16.1 INTRODUCTION

When both exposure assessment and the study of work-related musculo-skeletal diseases have highlighted a significant risk associated with repetitive movements of the upper limbs in various tasks, the need arises to implement specific measures aimed at re-designing jobs and procedures. Their efficacy depends on three types of simultaneous actions: structural modifications, organizational changes and personnel training. Whilst the structural measures are almost universally accepted and widely recommended, actions involving organizational changes do not always meet with a unanimous consensus, nor does the scientific literature provide concrete examples. Instead it merely supplies general and routine advice such as: 'reduce excessively high job frequencies' or introduce adequate breaks or job alternatives. This paper aims to provide some concrete experiences of organizational measures for re-designing jobs and preventing disorders caused by repetitive movements of the upper limbs.

16.2 ORGANIZATIONAL MEASURE

Measures typically involving changes to the job organization become necessary when it has been ascertained that jobs feature excessively frequent technical actions and/or inadequate functional recovery periods.

Measures designed to improve these two fundamental risk factors (frequency and recovery periods) can often interfere with 'productivity' and therefore are less readily accepted by employers.

The authors have recently acquired a considerable amount of practical experience which may supply useful suggestions for embarking on the organizational restructuring of specific working activities.

In a large metal-working factory featuring assembly lines, a significant prevalence of upper limb disorders was detected (Carpal Tunnel Syndrome, tendinitis, etc.) in most cases attributable to repetitive tasks performed with excessive frequency.

On the advice of the local health unit, the company asked the authors to carry out a detailed risk analysis in order to develop options for re-designing workstations more ergonomically.

The exposure assessment identified the following problem areas:

(a) high-frequency actions (38–40 technical actions per minute);
(b) in general, minimal use of force: force peaks of between four and five on the Borg scale were demanded by only a few actions that could be easily singled out during the job cycle. In almost all cases, the company quickly found specific solutions for bringing the use of force to within acceptable limits;
(c) posture seldom ‘extreme’ and therefore easily corrected by making some structural modifications to the workstation;
(d) recovery periods taken primarily for physiological reasons rather than for the purpose of alternating jobs. The daily schedule included two morning breaks (10 and 15 minutes respectively), a 30-minute lunch break, and one 10-minute afternoon break. One simple change involved optimizing the recovery periods: the total duration of the physiological breaks was already sufficient; by simply redistributing the breaks, it was possible to ensure adequate recovery periods, without altering their overall duration. The company undertook to redistribute the physiological breaks (35 minutes $= 10' + 15' + 10'$) so as to obtain four breaks (two in the morning and two in the afternoon).

In this case, the last problem that needed solving was the high frequency of the technical actions.

The first and most obvious intervention (most obvious for the ergonomist, but certainly not for the company!) was to reduce the pace of the task (with a consequent decline in ‘productivity’).

This solution was kept as a last resort, in the event that the frequency of the task could not be reduced by any other means.

After several meetings with the production engineers, safety officers and supervisors, this first objective was defined: to identify methods for reducing the number of technical actions required to complete a job cycle, without compromising output.

In other words, this meant *optimizing – in terms of quality and quantity – the technical actions needed to complete the cycle characterizing the task.*

The production engineers, particularly those directly involved in designing how tasks are carried out (e.g. in accordance with ‘Motion and Time’ methods) (Barnes 1968) were already quite proficient in seeking to optimize the actions that needed to be performed to complete a task. Their experience and research, however, was generally aimed at reducing the number of actions performed and shortening the duration of the task, thereby increasing the number of pieces produced.

Through valuable co-operation between the ergonomist and the production engineer, it was possible to use the fundamental experience of the engineer not to enhance 'productivity', but to improve working conditions and thus the health of the workers.

Having established, in agreement with the management and plant engineers, that the aim was to reduce the number of actions without reducing output, the first step was to identify the means and methods of achieving the objective.

The documents and the videotapes made for the exposure assessment (Colombini 1998) were analysed by a task force comprising the ergonomist, the production engineer and plant supervisor. Each task was revised several times, after critically examining the way the actions were performed.

In order to reduce the number of actions contained in a cycle, the procedure described hereunder and summed up in Table 16.1 was used.

Table 16.1 Brief recommendations for reducing the frequency of manual actions.

1 – AVOID USELESS ACTIONS	
	⇒ added by the worker
	⇒ due to technical flaws
	⇒ due to obsolete technologies
2 – PERFORM ACTIONS USING BOTH LIMBS	
3 – REDUCE REPETITION OF IDENTICAL ACTIONS BY	
	⇒ introducing pre-assembled parts
	⇒ introducing semi-automatic manufacturing systems
	⇒ replacing the process with hi-tech solutions
4 – REDUCE 'AUXILIARY' ACTIONS BY	
	⇒ creating intersections between conveyor line and
	⇒ operating area
LIT JOBS or increase the number of workers performing any given job	

16.3 THE ORGANIZATIONAL INTERVENTION

16.3.1 Procedures

1st phase – analysis of 'useless' technical actions: during this phase, it is decided whether all the technical actions are strictly necessary. In practice, this means: (a) detecting any 'useless' actions added by the operator: in this case, the operator must be trained to perform no more than the useful actions actually required to perform the task; (b) detecting whether any actions added by the operator are entirely arbitrary or in fact conceal a technical flaw. Once the technical flaw has been identified, it can be eliminated, and the number of actions can be reduced; (c) detecting 'obsolete' actions. In the course of time, assembly lines may undergo small changes to the machinery, rendering certain actions 'obsolete'.

2nd phase – analysis of upper extremity usage when performing technical actions: once all 'useless' actions have been eliminated, the next step is to optimize the distribution of the various actions between the two upper extremities. Workers often tend to favour their dominant limb. Simple low-precision actions may be performed equally by both limbs, thus reducing the frequency of actions performed by the dominant limb.

3rd phase – analysis of 'identical' technical actions: during this phase, it is seen whether workers are repeating identical actions for a significant portion of the job cycle. Together with the engineering staff, it was observed that the repetition of identical manual actions can often be avoided by introducing a specific mechanical device. When identical manual actions have been identified but no suitable tools can be introduced and, at the same time, the action frequency considerably increases the total frequency, one of the following solutions may be adopted: (a) eliminate the specific manufacturing step, by having the part arrive pre-assembled elsewhere; (b) introduce a semi-automatic step to replace the manual actions (high-cost solution); (c) re-examine the phase scientifically to find alternative solutions capable of fully by-passing the specific action sequence (hi-tech solution that often also improves the product).

4th phase – analysis of 'auxiliary' actions: it must be checked whether in passing from one cycle to the next, any 'auxiliary' actions are performed. It is generally useful to have the conveyor belt and operating areas cross each other in such a way as to avoid the worker having to pick up and replace pieces. It is equally helpful for the piece to reach the worker 'facing the right way' so as to minimize handling.

5th phase – when jobs need to be split: despite carefully reviewing actions, sometimes their frequency remains excessively high (up to even 60 actions/minute). In such cases, jobs need to be split.

16.3.2 Results

Although the authors realize that these recommendations are a start but are not exhaustive, the procedures described achieved in the specific context considerable improvements. The actions diminished in frequency from 38–40/minute (line average) to 30/minute: not only did the mean frequency levels decrease (in particular for the right hand), but the standard deviations also dropped dramatically: in other words, the most dangerous frequency peaks were eliminated.

16.4 CONCLUSION

It should be kept in mind that the frequency of 30 actions per minute – with no other risk factors involved is assumed to be the reference frequency (Occhipinti 1998). Several workstations still feature frequency levels higher than this value. In this case, it is necessary to at least introduce hourly job switches, so that the workers can alternate with less strenuous jobs.

In addition to alternating jobs in order to prevent disorders, the presence of adequate recovery periods is of critical importance. In another paper (Colombini 1998), the authors extensively report that rest periods can be optimized by ensuring a ratio of work periods to recovery periods of 5 : 1, within each hour of repetitive work. Often factories schedule long enough recovery periods but these are poorly distributed throughout the duration of the repetitive task.

The authors therefore suggest: (a) optimizing the distribution of official breaks: it is preferable to shorten each individual break, but to increase their frequency; (b) arranging, if possible, for rest periods to be scheduled at the end of an hour of repetitive work; (c) avoiding the scheduling of rest periods too close to meal breaks and to the end of shifts, in order to 'exploit' these as recovery periods; (d) rotating workers in non-repetitive tasks, so as to obtain an optimal distribution of repetitive and non-repetitive tasks, thus ensuring a good work/recovery period ratio.

16.5 REFERENCES

Colombini D. An observational method for classifying exposure to repetitive movements of the upper limbs. *Ergonomics* 1998; 41, 1290–1311.

Occhipinti E. OCRA: a concise index for the assessment of exposure to repetitive movements of the upper limbs. *Ergonomics* 1998; 41, 1312–1321.

Chapter 17

Regulatory Issues in Occupational Ergonomics

Giorgio A. Aresini

1. When the Treaty of Paris established the European Coal and Steel Community in 1951 with the obvious intention of helping to rebuild Europe after the devastating effects of the Second World War, it was considered vital to link technical and economic considerations with improving safety and health at work.

From 1954 onwards, the mining and steel industries and research institutions were therefore asked to look at these areas under the terms of specific programmes launched by the ECSC in order to investigate the factors which presented a risk to the safety and health of workers and to propose practical methods of prevention.

One of these programmes was devoted to research in the ergonomics sector with six consecutive five-year programmes being run, covering a total of 487 projects financed by the ECSC to the tune of €96 million.

Much of these research activities were dedicated to ergonomic analysis of work stations in order to elicit the findings needed to pave the way for proper correction of musculoskeletal disorders amongst workers. A large number of practical solutions were found for specific work stations in the two industrial sectors.

2. The experience acquired in these research programmes was a key factor in persuading the European Commission to propose in 1988 a Council Directive on the minimum safety and health requirements for the manual handling of loads where there is a risk particularly of back injury to workers.

This proposal became Council Directive 90/269/EEC in May 1990 (Official Journal of the European Communities L 156, 1990), the fourth individual directive within the meaning of Directive 89/391/EEC, which had created the legislative framework for improving health and safety for workers.

For the purposes of the Directive, manual handling of loads was understood to mean any transporting of loads including lifting, putting down, pushing, pulling and moving loads involving a risk particularly of back injury for workers.

The Directive required employers to seek to obviate the need for manual handling of loads by workers. They were to take proper administrative measures or, for example, use machinery. In cases where it was not possible to avoid manual handling, employers had to seek to reduce the risks. The onus was also upon them to tell workers how much the load weighed, where the centre of gravity of its heaviest part was and the risks involved in incorrect handling.

The factors for deciding whether there are general or specific risks for workers are listed in the annexes to the Directive. A load might, for example, present a specific risk of back injury if it is difficult to grip or if it is not stable and an activity might present a risk if it is too frequent or does not allow for sufficient rest.

All the Member States of the European Union have transposed the Directive into their domestic law.

3. In 1990 a Commission Recommendation (90/326/EEC) (Official Journal of the European Communities L 160, 1990) on the use of a European schedule of occupational illness was also adopted. It was recommended, *inter alia*, that the Member States introduce as soon as possible into their national laws, regulations or administrative provisions concerning scientifically recognized occupational diseases liable for compensation and subject to preventive measures the European schedule in Annex I which includes eight different groups of musculoskeletal disorders amongst the ailments caused by physical agents, namely:

1. osteoarticular diseases of the hands and wrists caused by mechanical vibration;
2. angioneurotic diseases caused by mechanical vibration;
3. diseases of the periarticular bursae due to pressure;
4. diseases due to overstraining of the tendon sheaths;
5. diseases due to overstraining of the peritendineum;
6. diseases due to overstraining of the muscular and tendinous insertions;
7. meniscus lesions following extended periods of work in a kneeling or squatting position;
8. paralysis of nerves due to pressure.

In order to provide medical information notices on the diagnosis for the items listed in Annex I of the above Recommendation, an expert group was convened by the European Commission.

These information notices, which were published in 1994 (Information notices on diagnosis of occupational diseases; Report EUR 14768, 1994), provide information pertaining to the causal relationships between diseases and exposures in the workplace. The notices constitute a source of information for interested parties (physicians, hygienists, social partners, national authorities, etc.), because it is clear that the methods for reporting, recognizing and paying compensation for occupational diseases in the various Member States are still far from uniform.

The notices are based on the published evidence obtained as a result of scientific investigations which have taken place over the last 30 years or more. These investigations contribute to the pool of knowledge on diseases, identifying competing causes including those which are occupational in nature. The expert group applied accepted scientific criteria when evaluating and selecting sources of information. Although scientific evidence does not always exist for high-exposure situations, the possibility of establishing a causal link in a specific case of disease should not be excluded.

Because of developments in technology and in scientific and medical methods over time, earlier results have had to be appropriately interpreted to take this into account.

In 1995, a total of 28,295 cases of occupational diseases caused by physical agents were recognized in the entire European Union, of which 2 539 were caused by mechanical vibration, 2,820 were osteoarticular diseases of the hands and wrists and 2,454 angioneurotic diseases caused by pressure, 2,305 diseases of the periarticular bursae and 3,392 paralysis of the nerves (provisional data from the EODS (European Occupational Diseases Statistics) project run by DGV in liaison with Eurostat, which are currently being validated).

4. Within the legislative framework, a proposal for a European Parliament and Council Directive on the minimum health and safety requirements regarding the exposure of workers to the risks arising from physical agents (vibration) (Official Journal of the European Communities C 77, 1993) is currently being examined by the Council.

Both hand–arm vibration and whole-body vibration are covered by the proposal: the first entails risks to the health and safety of workers, in particular vascular, bone or joint, neurological or muscular disorders, the latter entails risks, in particular lower back morbidity and trauma of the spine, as well as severe discomfort.

Discussions on this proposal will continue at the Council during the current Finnish Presidency possibly resulting in a Common Position before the end of 2000.

5. The assessment and limitation of risk is a vital element in maintaining or improving the minimum standards of protection which form the basis of a dynamic and efficient economy. If performed inadequately it can become a major cost both to employers and to workers and their families in financial, social and human terms.

Thus the reduction of risk must continue to be at the forefront of actions to improve the working situation.

In the framework of its Community programme 1996–2000 on Health and Safety at Work, the Commission pursues an investigative approach. It works closely with the European Foundation for the improvement of Living and Working Conditions and similarly with the European Agency for Safety and Health at Work. It continuously engages in dialogue with Member States, the social partners and the scientific communities. Through these sources the Commission will be able to identify areas where workers are not adequately protected by the existing legislative framework. These may include new high-risk activities, specific sectors of industry with unique problems or exceptional risks, categories of workers excluded from the present legislation.

Once high-risk activities have been identified, the Commission must then consider the most appropriate ways and means of combating them. In so doing the Commission will rely on a non-legislative alternative to provide an effective solution. However, if the legislative solution is deemed the most appropriate in the interests of the health and safety of the workers concerned, the Commission will act accordingly.

6. Musculoskeletal disorders have long been a major cause of suffering in many industrialized countries. Besides the lower back region, the neck and upper extremities are the regions affected most.

For these reasons the European Commission (Directorate-General V) has requested the assistance of the European Agency for Safety and Health at Work to conduct preparatory work on the risks factors for work-related neck and upper limb musculoskeletal disorders (WRULDs).

This work would enable the Commission to reply to several requests from social partners and the European Parliament to address this problem as a key issue within the next work programme. In this context it would be possible to take a position before the European Parliament's Social Affairs Committee which stated that 'Guidance is not enough and legal obligations must be placed on employers to prevent WRULD'. Without prejudice to the kind of Community action involved, this analysis would be the main basis for it.

The European Agency invited Professor Peter Buckle and Dr Jason Devereux from the Robens Centre for Health Ergonomics at University of Surrey, UK, to prepare a report containing the state of the art and the scientific knowledge on the relationship between work and neck and upper limb disorders.

Some of the considerations and conclusions (paragraph 7 to 12) are taken from the above draft report.

7. During the course of their study, the authors attempted to provide answers to the following questions:

- What is the extent of the problem within Member States of the European Union? Is there a need for Community-level action? Would such action create added value?
- Might epidemiological evidence suggest a course of action to avoid risk factors?
- Is there coherent supporting evidence from the scientific literature on the mechanisms that might cause these disorders?
- Does intervention in the workplace reduce the risk?
- What preventive strategies might be recommended?

The necessary information was collected from a careful review of existing literature, followed by a meeting of experts in The Netherlands in October 1998, and wide consultation with further experts and interested parties.

8. One of the first problems the authors faced was the availability of diagnostic criteria. Analysis of the criteria developed in various EU Member States led to the conclusion that there was no common approach. Although the authors recognize the importance of these criteria, they stress that the strategies for prevention that they recommend for avoiding the risk of such disorders occurring are clearly not dependent on diagnostic classifications.

The relationship between symptoms, the reporting of accidents, injuries and disablement remains unclear.

9. Basing themselves on numerous national-level studies on the incidence of such disorders, the authors confirm that WRULDs are a significant problem in terms of ill health and associated costs within the workplace. Moreover, it appears quite likely that the scale of the problem will increase, given the rise in exposure to risk factors.

Unfortunately, as stated above, the lack of standardized diagnostic criteria makes it difficult to compare data between the various Member States. A solution to this problem may be provided by work currently being done on the compilation of statistics on occupational diseases (see paragraph 3).

What is more, even those studies which used the same methodology showed significant differences between the Member States concerned as regards the rate of incidence and the assessment of costs incurred.

The authors would like to emphasize that all players in this field clearly agree that action should be taken.

10. Much of the report is devoted to an understanding of the pathogenesis of these occurrences, which varies significantly according to the type of the disorder. For many of them (and carpal tunnel syndrome is a prime example), the body of knowledge is impressive, providing information on biomechanics or on the measurement of soft tissue changes. Yet, information in other areas is very limited and can only be used to form working hypotheses.

11. Scientific literature clearly demonstrates the causal links between risk factors in the workplace and many disorders of this nature. That the causes of some pathologies remain unidentified indicates that related studies are in short supply.

Confirmed links exist as regards:

- the high incidence of disorders among workers exposed to high biomechanical loads;
- the effect of exposure to vibration on the arm, forearm and hand;
- the role of various factors, such as posture, force applications at the hand, direct mechanical pressure on tissues as well as cold.

Clearly, the interaction between the various factors is often difficult to interpret. Nonetheless, it must always be taken into consideration.

It is also worth again calling to mind that a proposal for a Community Directive (see paragraph 4) aimed at reducing the harmful effects of vibration on workers is under discussion. It will provide guidelines on a number of specific exposure factors. This proposal is the first stage of a wider strategy on a whole range of physical agents and may be adopted as a Directive by the European Parliament and the Council in 2001.

12. The last part of the report sets out the principles that justify preventive action. As is the case for all other risk factors at work, any action must be based on an accurate risk assessment. In turn, this should reveal what preventive measures and health surveillance programmes are needed.

For this reason, the report states which factors should be taken into consideration. As stated above, these are posture, force applications at the hand, exposure to vibration, direct mechanical pressure on tissues, effects of a cold work environment, and a series of variables pertaining to the way work is organized, including psychosocial issues.

For practical purposes, the authors have set four hours of exposure as the first parameter indicating the need for preventive action.

Next, they identify the various body parts (neck, shoulder, wrist) whose physiology and pathology need to be analysed in order to assess the factors responsible for the disorder.

The authors consider current knowledge on the effectiveness of various strategies to be incomplete. Nonetheless, a systematic, ergonomic approach that is based on known risk factors must clearly be the first choice.

The authors emphasize that appropriate ergonomic action taken to solve one problem often has beneficial effects on related disorders. They give the example of reducing the exposure of hands to vibration. Besides reducing the risk of Raynaud's syndrome developing, this also reduces the risk of tendinitis in the hand or wrist, which is linked to the need to use more force.

In conclusion, the authors stress the importance of a health and risk surveillance programme, which they see as an additional, but significant, factor to be considered when deciding what action is to be taken.

13. As mentioned above, it is not currently possible to predict what action the European Commission will take in this field, which is without doubt one of the emerging problem areas in relation to disorders (or at least of damage) associated with work activities.

The Commission's departments, in conjunction with the tripartite Advisory Committee for Safety, Hygiene and Health Protection at Work, must carefully analyse the information provided by the European Agency's report, particularly the section addressing the justification for Community-level action. The first meeting of an ad-hoc group on the subject is already scheduled for the beginning of October. The Committee should therefore be able to express its opinion on what action to take and how to proceed during the plenary session in the spring of 2001.

Taking the European Parliament's position into account, the most attractive option might be to draft a proposal for a directive, a binding instrument for EU Member States. However, although such a choice is completely justifiable, it might conflict with Member States' general policies, as they prefer to limit Community legislation in the field of health and safety at work to what currently exists, and concentrate on applying it correctly.

In any case, the European Commission, given its own work programme as described in paragraph 5, certainly does not intend to forgo its prerogative, in other words its right of initiative to propose the instruments it considers most useful for the harmonious development of Community policy.

It will become easier to assess this in the next few months, once the new Prodi Commission has had time to define its own priorities more clearly. Its first meeting is planned for 22 September.

REFERENCES

Information notices on diagnosis of occupational diseases, Report EUR 14768. Office for Official Publications of the European Communities, 1994.

Official Journal of the European Communities L 156 of 21.6.90, page 9.

Official Journal of the European Communities L 160 of 26.6.90, page 39.

Official Journal of the European Communities C 77 of 18.3.93, page 12.

Index